New Ice Age Near:
58-Part Evidence

by Rolf A. F. Witzsche

Contents

About the Illustrated Science series: 10

*New Ice Age Near - 58-part evidence 12

The Sky is Falling 13

A large body of evidence exists 14

Refer to the transcript pages 15

A more-orderly summary 16

*Paradoxes in Ice Age perception 17

Evidence measured in historic ice deposits 18

Colder than the Little Ice Age 19

Enormous climate fluctuations 20

Evidence is laid up in the great ice sheets 21

In deep-sea sediments 22

Doctrine of the Constant Sun 23

Ptolemy's exotic theories in astronomy 24

The Milankovitch Cycles theory 25

Forged by the box of doctrines 26

Milankovitch Cycles 27

It doesn't match up 28

In spite of the elegance in reasoning 29

Main arguments for the orbital cycles theory 30

Ice sheets more than 10,000 feet deep 31

*The Astrophysical Paradoxes 33

To match the Ice Age evidence 34

Internally-powered Sun theory 35

The slow heat-transfer theory 36

In the electric universe 37

Researcher David LaPoint 38

At the Los Alamos National Laboratory 39

A magnetic confinement dome 40

Facilitates the concentration of plasma................................. 41

David LaPoint discovered 42

The Primer Fields principle 43

Heating in the surface of the Sun 44

The Sun delivers proof 45

The Sun takes the process no further 46

In the electric Sun model 47

Matches elements in the cosmic abundance table 48

The Sun is an electric powered star 49

The Big Bang model, is dangerously entropic......................... 50

The red-shift phenomenon 51

The red-shift becomes understood....................................... 52

The photons, which are the carriers of light.......................... 53

A moving electromagnetic wave.. 54

The high-energy fields.. 55

Different wavelengths.. 56

When light is propagated over long distances........................ 57

The red shift invalidates the Big Bang theory 58

Liberation from the Big Bang Doctrine of Entropy 60

Humanity is definitely anti-entropic 61

The term, theory, no longer applies 62

Big Bang theory illustrates exotic excuses............................. 63

In this context of the empty box... 64

Sun in a still-ongoing dynamic process 65

The alpha fusion chain 66

The distribution pattern proves... 67

4

Ratio of atomic elements 69

Atomic elements flowing away from the sun 70

Distribution of synthesized atomic elements 71

In entropic science 72

Enormously large clouds of synthesized elements 73

Synthesized giant clouds also reveal...................................74

The double-bowl geometry of the Primer Fields 75

Plasma inflow geometry has been physically verified 76

The phenomenon of the noctilucent clouds.........................77

The rings of Saturn 78

Saturn's moon, Dione...79

Saturn proves with its moons 80

Looking through the umbra of the sunspots 81

The Sun a plasma sphere 82

Stars with a mass 150 times greater 83

A simple size comparison...84

Concepts of the entropic star cycle......................................86

The wide scene of Ice Age historic evidence 87

Cosmic universe is ruled by a much greater force 88

The ecliptic phenomenon is expected 89

Saturn's rings are made of water ice 90

The ecliptic phenomenon 91

The thin disk of the galactic plane 92

Two plasma confinement domes of the Primer Fields 93

Exceedingly puzzling in entropic cosmology.........................94

The moons of the planets 95

The rings of Jupiter are fainter...96

The rings of Uranus 97

The rings of Neptune are still fainter 98

Neptune is the most distant planet 99

The coldest object found in the solar system 100

Supersonic air movement 101

By the principle of magneto hydrodynamics 102

Something makes Saturn special 103

Saturn's exceptional storms.................................. 104

Saturn's exceptionally strong polar aurora 105

Aurora phenomenon on Jupiter.............................. 106

Jupiter's storms are nevertheless gigantic..................... 107

Jupiter is a cold planet 108

Aurora on Earth are relatively weak 109

Perpetual motion machines are not possible 110

In the Cat's Eye nebula..................................... 112

Some light nuclear fusion synthesis........................... 114

Researcher David LaPoint regards a nebula 115

The famous Crab nebula 116

The entire galaxy is in motion. 117

The plasma streams in the galaxies 118

Stars cannot explode....................................... 119

A puff of smoke 120

Smoke from the supernova 1987A 121

Light is slowed by denser media 122

A tell-tale image 124

Electric-universe supernova events........................... 125

The Crab a pulsar 127

A neutron star is physically impossible 128

A pulsating system that generates high-energy bursts 130

The apparent star in the Crab 131

In the Eagle Nebula 132

Named the Pillars of Creation 133

The gigantic features are barely recognized 134

At the focal point of a single large Primer Field 135

In the electric universe.. 136

Intergalactic plasma streams 137

Globular star clusters in the halo 138

Spherical amalgamation of stars.. 139

Globular star clusters 140

Intergalactic plasma streams resonance.. 141

Between the Milky Way and Andromeda 142

The longer and more dominant climate cycle 143

The combination of two resonance effects 144

The entire development of mankind 145

Our Sun is a rather mediocre one 146

Through the last glaciation period 147

Plasma streams between the stars Sirius and Vega............................. 148

*Critical evidence of a changing world 150

Dense plasma sphere surrounding the Sun 151

Primer Fields static experiments.. 152

High-energy electric flow experiments... 153

Revealed the existence of a magnetic confinement dome................... 154

Magnetic retention strength.. 155

Excess plasma becomes solar wind.. 156

venting' feature 157

Concentrated in the fusion reaction cells 158

Electrically powered fusion generates some thermal heating 159

For as long as the solar winds flow 160

The occurrence of sunspots is proof...................................... 161

The measure of the solar-wind pressure 162

NASA's Ulysses measured a 30% reduction 163

Magnetic measurements 164

The dynamo effect of the spinning Earth 165

Magnetic effect from the Primer Fields................................ 166

When the Primer Fields are strongly dominant.................... 167

Zero degrees' deflection would mean 168

A shift of the boundary zone 169

Diminishing magnetic-pole deflection 170

Maintaining the spin-rate of the planets 171

Ptolemy had bent to the doctrine.. 172

Hannes Alfven model of the electric universe 173

How soon the Sun will go inactive 174

Absence of tell-tale boundary conditions............................ 175

Vulnerable to a sudden collapse without warning 176

A reactive response is not sufficient 177

Experience a 70% loss of solar energy 178

Highly-efficient self-regulating features 179

Diminishing that Ulysses has reported 180

Resulting from the increased cosmic-ray-flux 181

We may not even have 30 years left 182

Electric weakening far-reaching reflections........................ 183

Might give us potentially 30 years 184

White Dwarf examples... 185

A White Dwarf is a star that became inactive...................... 187

The Sun had recovered for brief periods 188

Dust accumulations in glacial deposits 189

Dust in ice, and White Dwarf stars... 191

The entire world is affected 192

Free universal housing in new cities 194

In honour of Canada's birthday celebration 196

American birthday celebration on the Fourth of July.............................. 197

Supporting exploration videos.. 198

Celebrating the Near New Ice Age 199

Cold Fusion powers the sun 200

Ice Age of the dimmer sun in 30 years.. 201

Our electric fusion Sun... 202

Electric planets electric Earth.. 203

Electric Mars.. 204

Energy future... 205

Unlimited fresh water 206

About the Illustrated Science series:
Ice Age – Climate Change
and the book
Ice Age 58-part Evidence

Numerous fields of evidence tell us that the next Ice Age is near. Most of the evidence was discovered in the 1990s and thereafter. Some evidence is measured in ice cores; some is measured in space, by satellites. Some measurements are also made on the ground in terms of measurements of the Earth's magnetic-pole drift observed in northern Canada. All of this is seen combined with high-energy physics experiments at a leading national laboratory, and is also explored in the small in static experiments.

Against the background of these widely diverse types of evidence that have been recently discovered, the historic Little Ice Age in the 1600s, takes on a new dimension as a yardstick for measuring the future that by this evidence promises to be up to 40-times colder than the Little Ice Age had been. The evidence poses a challenge. Are we ready to respond?

In the Little Ice Age between 10% and up to 30% of the populations in Europe had perished by starvation. The last Big Ice Age was evidently vastly harsher. Only 1-10 million people emerged from it alive. That's all we had after 2 million years of development. We want to do far better this time around; and we can, with large-scale technological infrastructures for our food supply. But will we create them? Will we get the job done in the 30 years that we still have left before the Ice Age starts anew? And how certain are we that the phase shift to the next glaciation period will begin in the 2050s? We have 58 items of evidence to support this as a possibility. But will we move with the evidence?

It takes an independent researcher to brake the taboos that have kept mainstream cosmology imprisoned, increasingly, during the past century, even while what is regarded as taboo is known to be wrong.

The Illustrated Science series is intended to open the scene beyond the threshold of accepted taboos, to where the actual physical evidence speaks for itself.

The scope of the existential challenge that the Ice Age brings with it, takes astrophysics out of the academic domain and places it into the foreground as one of the most-critical issues of our time. The big Climate Change events that have already worldwide effects are mere fringe effects in the flow of the ever-changing cosmic dynamics. The big effect, when the Ice Age begins anew, promises to be caused by a dimmer and colder Sun with 70% less radiated energy. This defines our climate future.

Sure, we can live with all that by creating new platforms for agriculture that are able to operate under Ice Age conditions. But will we do it? The task is enormous. Or will we fail ourselves on this front? We have no reason to allow us to fail. We have the materials and energy resources on hand to accomplish everything that is required for us to continue to live in an Ice Age World. But will we do it? The big question that never goes away, therefore, is; will we develop our inner resources as human beings sufficiently to get the job done, and to get it done in time? Or will we do nothing, ignore the challenge, and condemn our children and one-another to an agonizing death by starvation? That's the choice.

Towards meeting the inner challenge, I have created the epic series of novels, The Lodging for the Rose. And further, towards meeting the science challenge, I have produced numerous research books and several dozen exploration videos that the Illustrated Science series is modeled after. The work is the result of a quarter century of research, for which numerous elements of evidence in related fields came to light during the timeframe of my research.

It is my hope that the work that went into all of these projects will help in some degree - for humanity that we are all a part of - to write itself a ticket to have a future.

High-resolution color images, of the images in this book, can be obtained at www.iceagetheatre.ca

New Ice Age Near - 58-part evidence
(Comprehensive summery of evidence that we live in an electric universe in which ice ages happen and the next one starts in 30 years under an electric Sun becoming inactive.)

The Sky is Falling

The original title was to be: The Sky is Falling. The title was meant to be shocking. When the general perception of our world lies so far off from what is actually real, so that reality becomes unbelievable, a shocking title would be warranted.

Of course the sky isn't falling, but the global temperature will be falling sharply when the ice age glaciation climate begins anew that has been the normal climate on earth for 85% of the last 2 million years.

A large body of evidence exists

A large body of evidence exists, that when it is seen as a whole in the context of causative principles, the composite indicates that the resumption of the glaciation climate will likely occur in the 2050s timeframe or sooner, and that the transition will be swift, possibly in the order of hours, days, or weeks, as the Sun goes 'inactive' with extremely dramatic consequences for all of humanity.

Refer to the transcript pages

transcripts at:
www.ice-age-ahead-iaa.ca

Photo taken by
Andrew Mandemake

Because the evidence is extensive, and concerns numerous aspects
of science, this video has been created as a comprehensive
summary. You may wish to refer to the transcript pages for this
video for references to the more detailed explorations involved.

A more-orderly summary

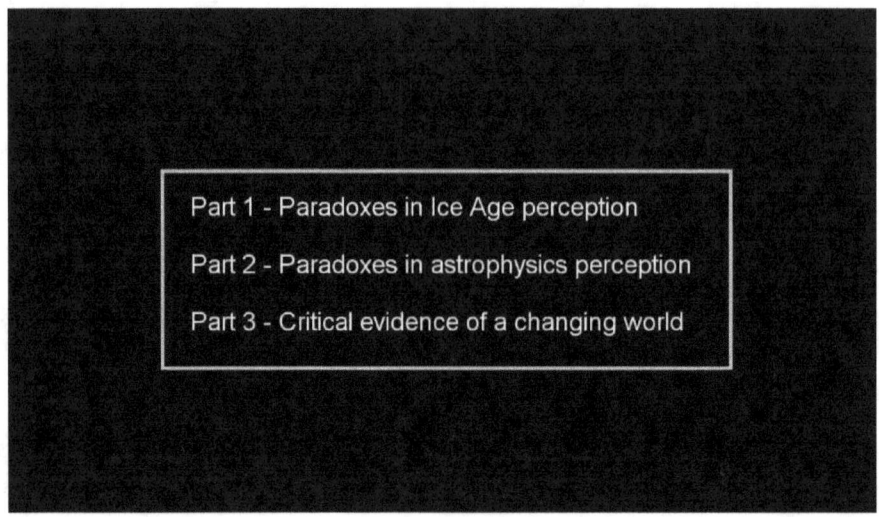

Part 1 - Paradoxes in Ice Age perception

Part 2 - Paradoxes in astrophysics perception

Part 3 - Critical evidence of a changing world

For a more-orderly summary, the video is divided into three parts.
Part 1 deals with paradoxes in general perception about the Ice Age phenomenon
Part 2 deals with astrophysical aspects that resolve the paradoxes
Part 3 deals with critical supporting evidence of the weakening solar system

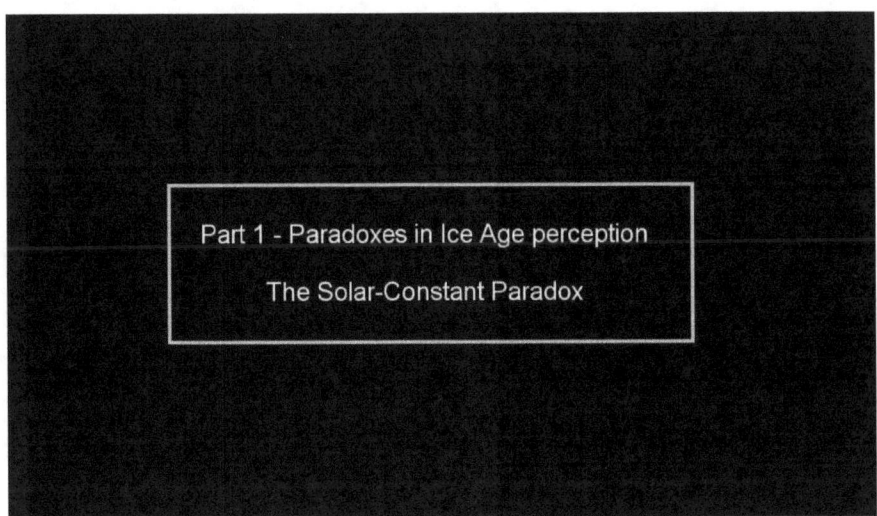

Part 1 - Paradoxes in Ice Age perception
The Solar-Constant Paradox.
The general perception is, that the Sun is an invariable constant for all climate considerations. The evidence indicates that this perception is not correct.

Evidence measured in historic ice deposits

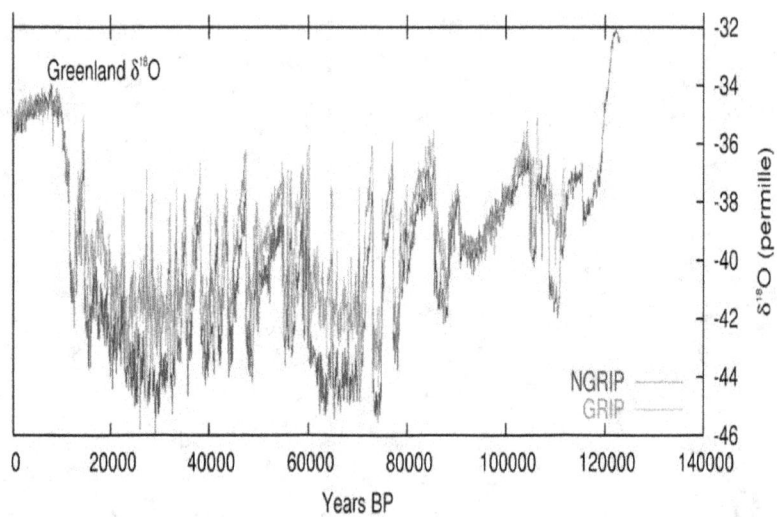

#1

We have evidence measured in historic ice deposits, of enormously fast global climate fluctuations that have occurred in the past, which are totally incompatible with the theory of a constant Sun.

Colder than the Little Ice Age

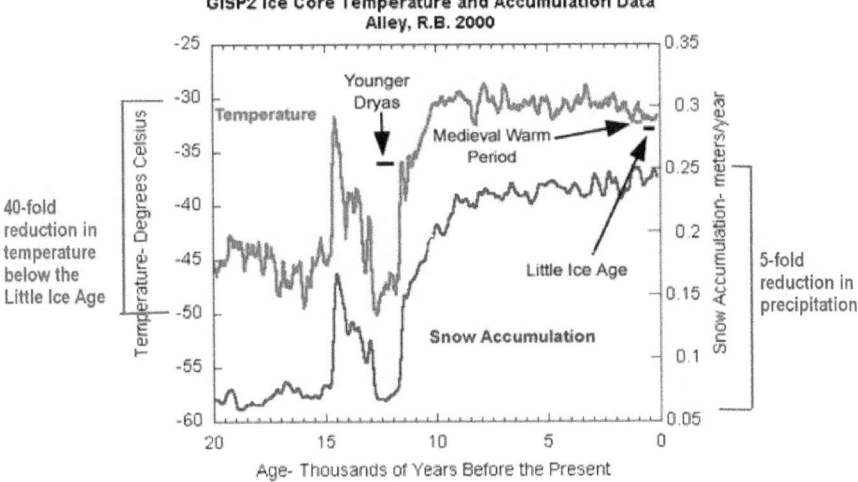

#2

We have measured cold climates that were 40 to 50 times colder than the Little Ice Age had been, with fast transition times between them.

Enormous climate fluctuations

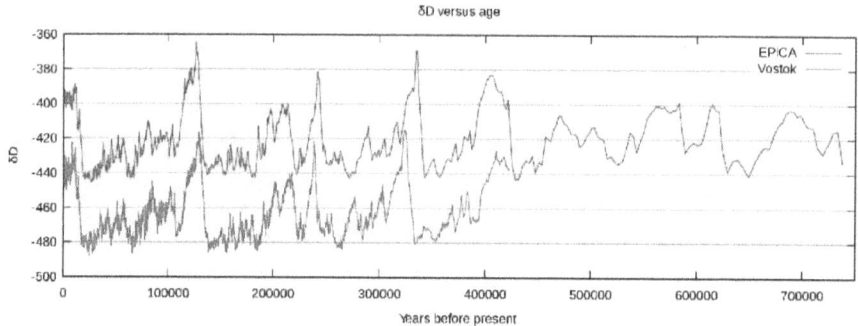

#3

We also have measured evidence on hand, that enormous climate fluctuations have occurred all the way through the last half million years that we have evidence for from different exploration sites.

Evidence is laid up in the great ice sheets

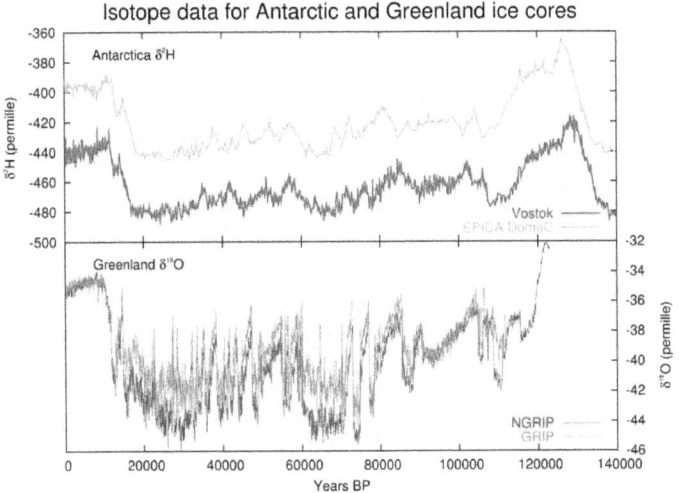

Isotope data for Antarctic and Greenland ice cores

The evidence is laid up in the great ice sheets in Antarctica, and in the ice of Greenland. Ice core samples have been examined from several drilling sites, both in Antarctica and in Greenland. The measurements from these widely separated sites all agree, within the limits of their resolution of details respective of the ice conditions. In considering the universal agreement of the measurements, one can assume that the measurements are essentially correct, even though, what they tell us, turns the conventional perception of solar dynamics upside down.

In deep-sea sediments

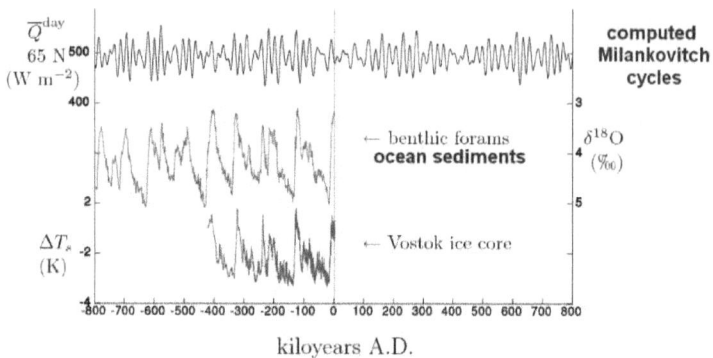

comparison of measurements between
ice core data and ocean sediments

The same evidence that is found in ice cores, is even found in deep-sea sediments. The results all tell the same story. They tell a story for which no real theory exists, since it is assumed by a fundamental doctrine that our Sun is an invariable star with a constant energy output. In other words, the evidence and the theory do not agree by a long way. For a while the paradox was all blamed on orbital variations. However, the computed results of the variations, don't agree with the evidence either. No part of the measured evidence agrees with any theory, since all science perceptions are constricted by the Doctrine of the Constant Sun.

Doctrine of the Constant Sun

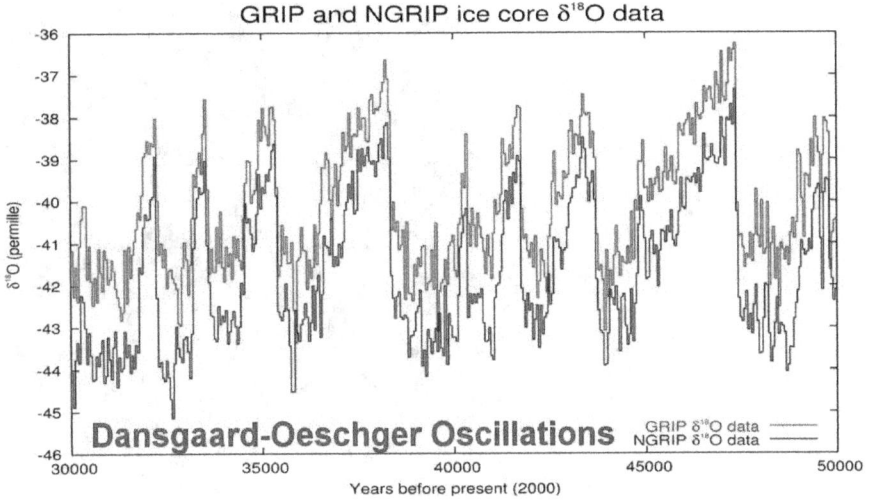

#4

In some cases the evidence tells us that the Earth warmed up from deep glacial conditions to near interglacial conditions in the space of just three decades. More than 20 such events have been detected in the ice deposits in Greenland from the last glacial period going back in time roughly 120,000 years. Numerous exotic theories have been invented to bring the measured evidence into conformity with the, Doctrine of the Constant Sun, that is widely accepted.

Ptolemy's exotic theories in astronomy

Ptolemy
being guided by philosphy

Margarita Philosophica by Gregor Reisch, published in 1508

However, when one examines the modern climate theory based on the constant Sun, one finds a striking similarity on all cases in this theory, with Ptolemy's exotic theories in astronomy, in adherence to doctrines, of epicycles and so forth, with which Ptolemy delivered some amazingly elegant 'scientific' proof of something that doesn't actually exist.

The Milankovitch Cycles theory

Milutin Milankovich

To solve the paradox, the orbital cycles theory was invented, named the Milankovitch Cycles theory, according to its main contributor. The theory is an elegant exotic example of climate theories that are constricted from the outset by the Doctrine of the Constant Sun. When one is boxed in like this, one has little to work with that is based on the reality outside the doctrine. That's the type of box that Ptolemy was restricted to in astronomy. He achieved marvels in this box, and so did Milankovitch whose achievement rivals in quality Ptolemy's work.

Forged by the box of doctrines

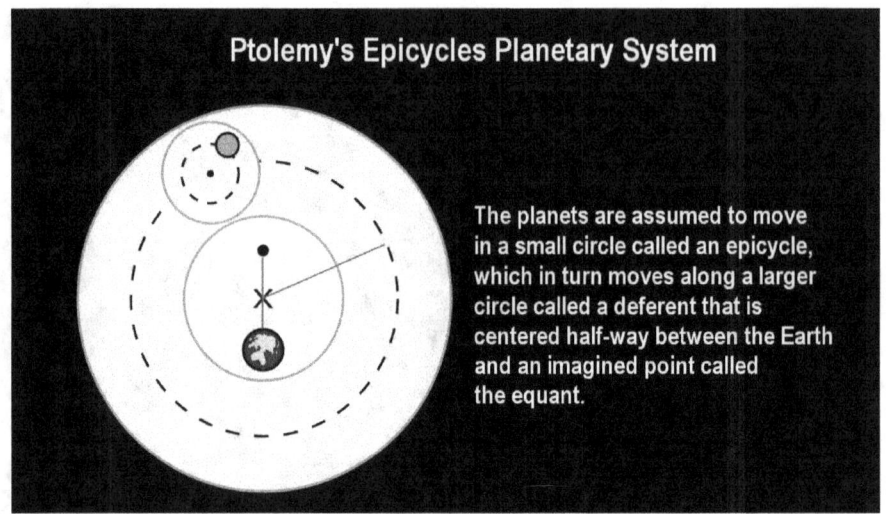

Ptolemy's Epicycles Planetary System

The planets are assumed to move in a small circle called an epicycle, which in turn moves along a larger circle called a deferent that is centered half-way between the Earth and an imagined point called the equant.

Ptolemy's calculations, forged by the box of doctrines, came extremely close to agreeing with the actual physical measurements, though they came not close enough.

Milankovitch Cycles

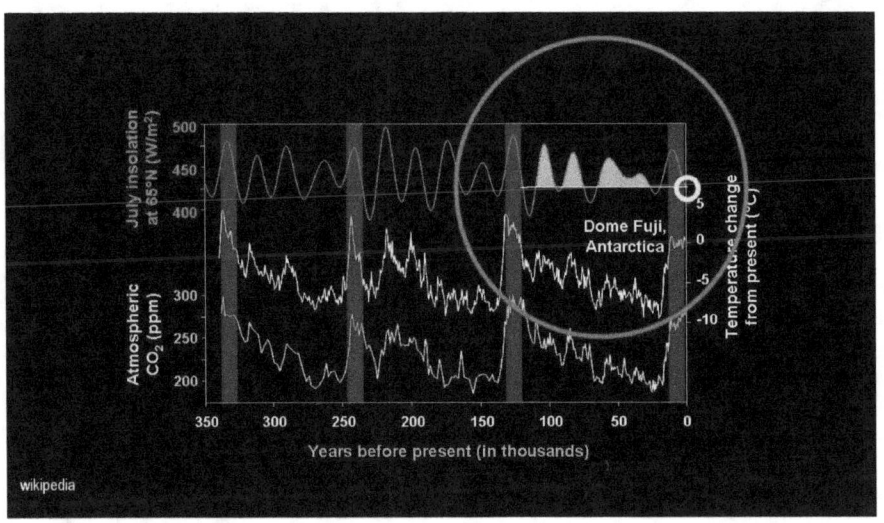

#5

The calculations of the Milankovitch Cycles are actually further off the track from the physically measured reality than Ptolemy was.

It doesn't match up

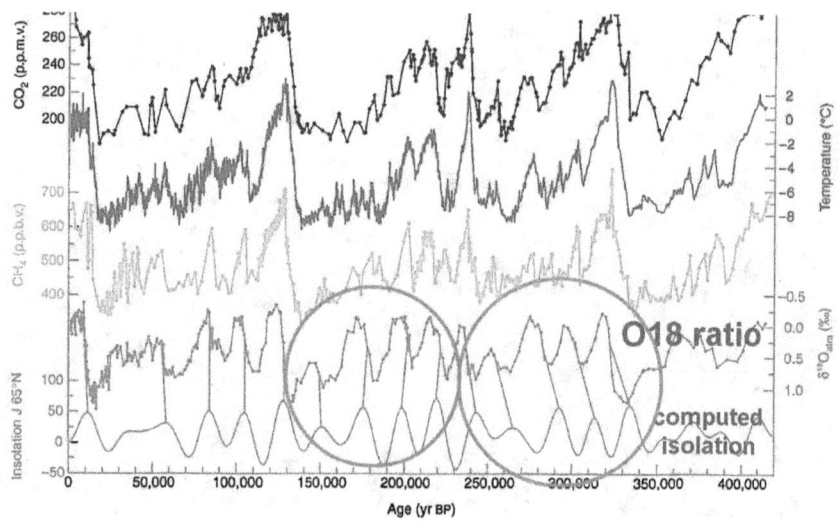

Of course, in the time of Milankovitch, the ice core historic data did not exist, for comparison. But this data exists now, and it doesn't match up by a long way.

Sometimes the measured results precede the computed, theorized cause by tens of thousands of years, and sometimes they lag by just as long. This means that greater forces are at work to produce the ice age cycles that are not mechanistically definable, but which are so large that the orbital cycles become consequential to them.

In spite of the elegance in reasoning

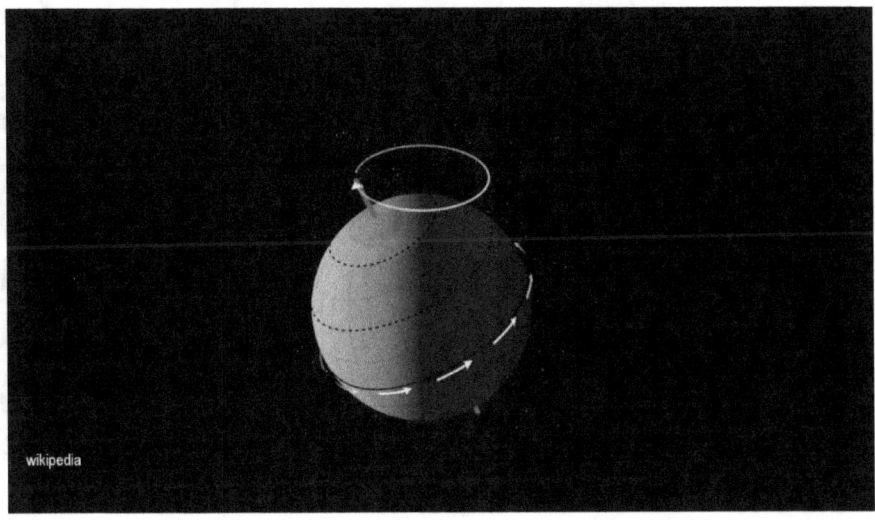

wikipedia

#6

In spite of the elegance in reasoning to correlate the celestial dynamics with what is in the box of the mechanistic, which on the surface actually makes sense to some degree, when seen from within the box, the orbital cycles theory ultimately fails, because it fails to consider that the total exposure of the Earth to the Sun remains always the same regardless of the cyclical variations of the tilt of the spin axis and its orientation, and the eccentricity of the Earth's orbit, which together only affect seasonal differences and hemispheric differences, but are never expressed as global differences.

Main arguments for the orbital cycles theory

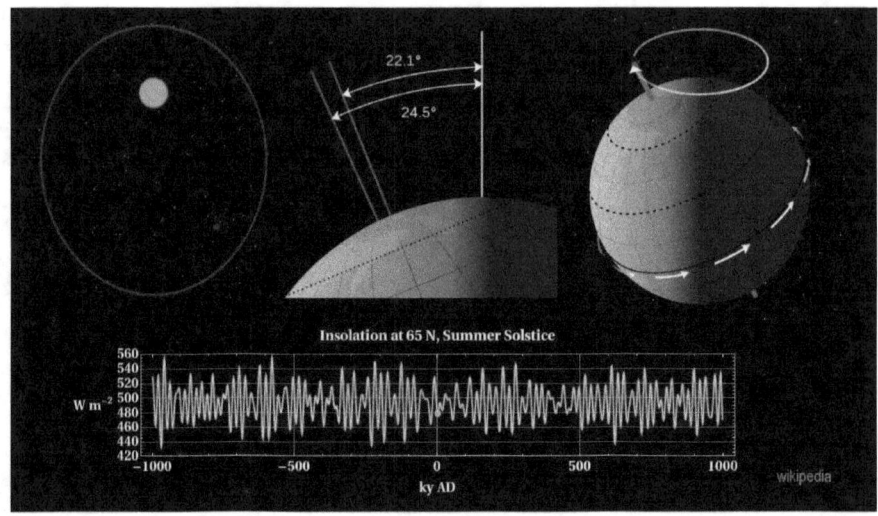

#7

One of the main arguments for the orbital cycles theory is, that an Ice Age may occur as the result of extreme seasonal and hemispheric distribution of the solar energy received, even though the total energy received remains the same. However, no measured evidence exists that hemispheric climate shifting has any significant effect, much less the gigantic effects that can cause the enormous ice ages that we have evidence of. All measured evidence indicates that the Ice Age effects, when they do occur, are simultaneously global in all respects.

Ice sheets more than 10,000 feet deep

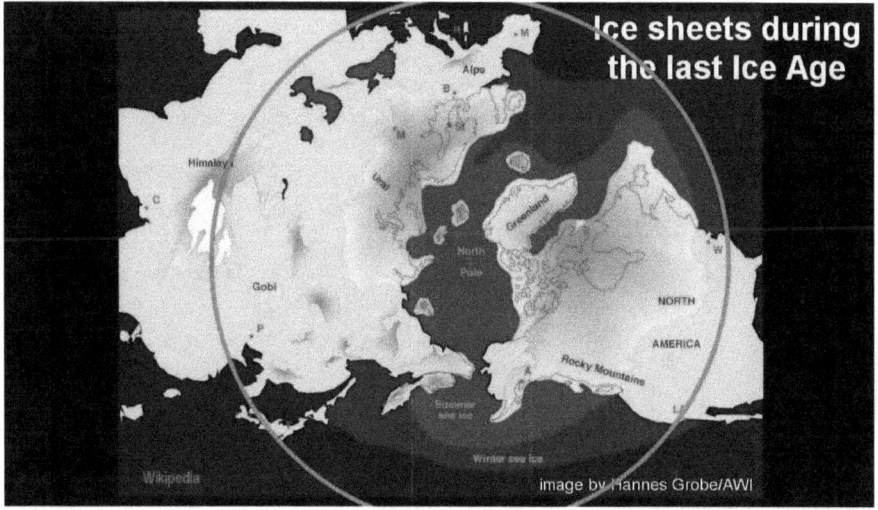

#8

All of this put together means that no rational theory actually exists in support of the various types of Ice Age evidence, including the fact that during the last glaciation condition the ocean level became lowered by 400 feet as evaporated water was piled up on land in the form of ice sheets more than 10,000 feet deep, and that permafrost had reached as far south as Beijing, and the winter sea ice had reached as far as Los Angeles.

These enormously severe climate conditions do not occur without a correspondingly large cause. And this enormously large and ever changing cause is fundamentally inconsistent with the Theory of the Constant Sun, which is the current theory that rules astrophysical science.

However, the observed and measured evidence, and its numerous corroborating types of evidence of enormous climate events, do make all perfect sense suddenly, the moment that one breaks through the barrier erected by the Doctrine of the Constant Sun. When this breakthrough is made, the Ice Age paradoxes all become resolved.

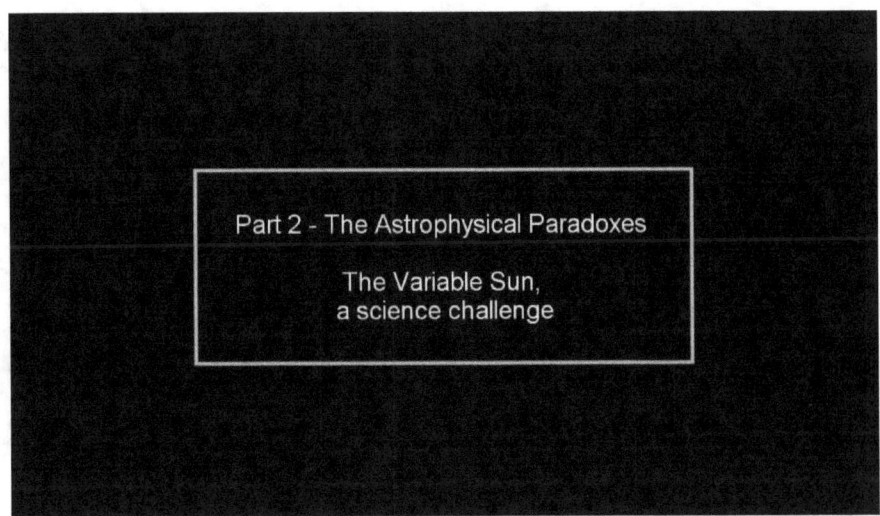

Part 2 - The Astrophysical Paradoxes

The Variable Sun appears to be the greatest science challenge of our time.

To match the Ice Age evidence

The Sun at
a plasma node

David LaPoint - The Primer Fields

The only type of variable Sun that is physically possible within the framework of the known or knowable physical principles, and is extensive enough in its expression to match the Ice Age evidence, is the electric Sun.

This sun is electrically powered within a sphere of strongly concentrated interstellar plasma, that is focused on it electromagnetically by a system of self-forming Primer Fields. Such a sun, being electrically powered with nuclear fusion occurring on its surface, can meet all the requirements for causing the ice ages, and related evidence.

Internally-powered Sun theory

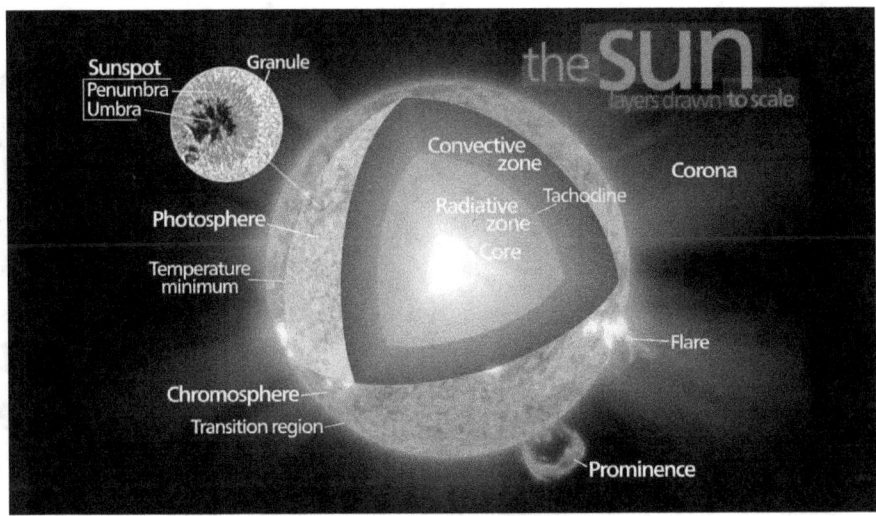

In comparison, the widely accepted theory of the Sun, the internally-powered Sun theory, actually affirms that it is bankrupt on this account, by stating emphatically that the internally powered Sun is a constant factor that simply cannot change.

It is theorized that it takes up to 30 million years for heat flux from this Sun's core to work its way to the surface, or up to 170,000 years for the fast thermal flux in the form of photons to cross the 700,000 kilometre distance from the center of this Sun, to the surface.

The slow heat-transfer theory

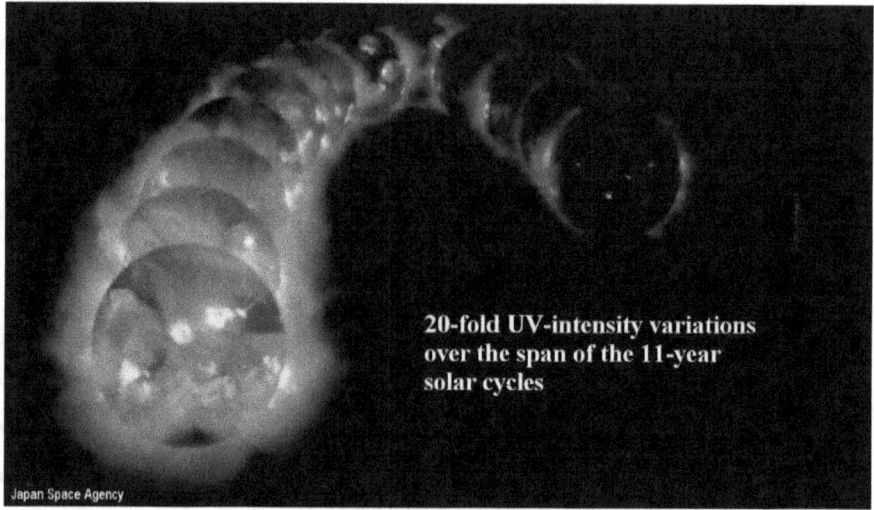

20-fold UV-intensity variations over the span of the 11-year solar cycles

Japan Space Agency

#9

The slow heat-transfer theory, of course, affords no basis for even the well known 11 years solar-activity cycles to exist, during which the Sun's energy output in the high UV band varies by a factor of 20. This variance has been measured, even photographed. What we see here is a totally natural phenomenon for an externally powered electric Sun, but an impossibility for an internally heated Sun with a slow oozing heat transfer mechanism. The evidence that we see here is typical for an electric Sun with surface reactions that rapidly respond to resonating external electric conditions.

Electromagnetic processes are inherently variable processes. The intensity of the expression of the operating principles, always varies with the density of the input energy that is driving the processes.

In the electric universe

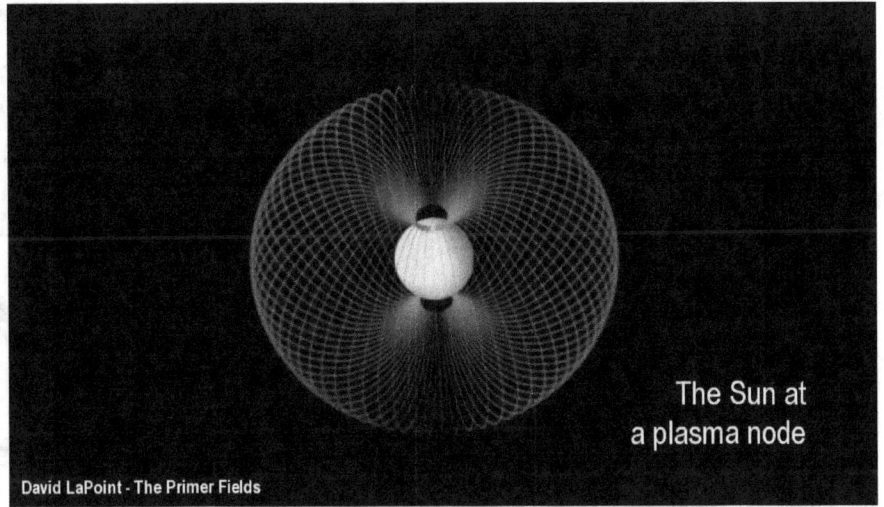

The Sun at
a plasma node

David LaPoint - The Primer Fields

In the electric universe, the Sun can be seen as existing within a large electromagnetic structure that concentrates interstellar plasma streams and focuses the pumped-up plasma onto the Sun. The process results into a dense plasma sphere surrounding the Sun.

Researcher David LaPoint

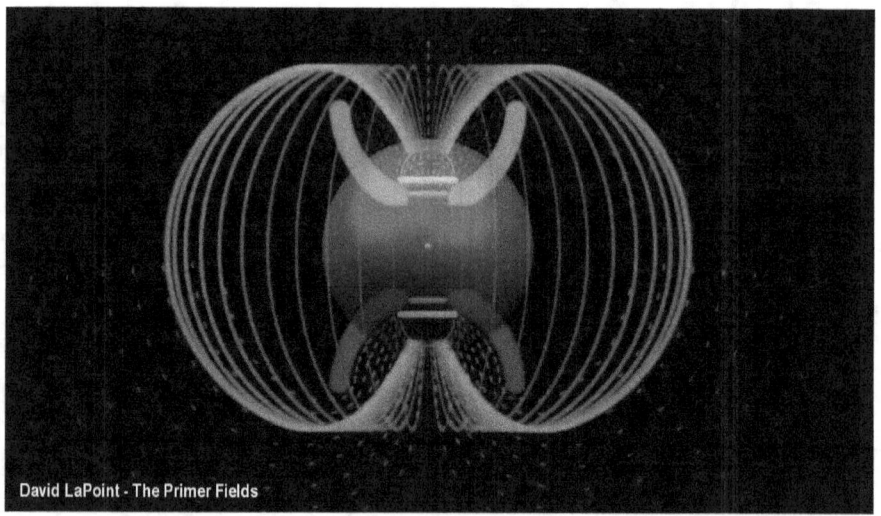

David LaPoint - The Primer Fields

The dynamics of the electromagnetic structure has been explored in detail by the researcher David LaPoint, who has termed the structure, the Primer Fields.

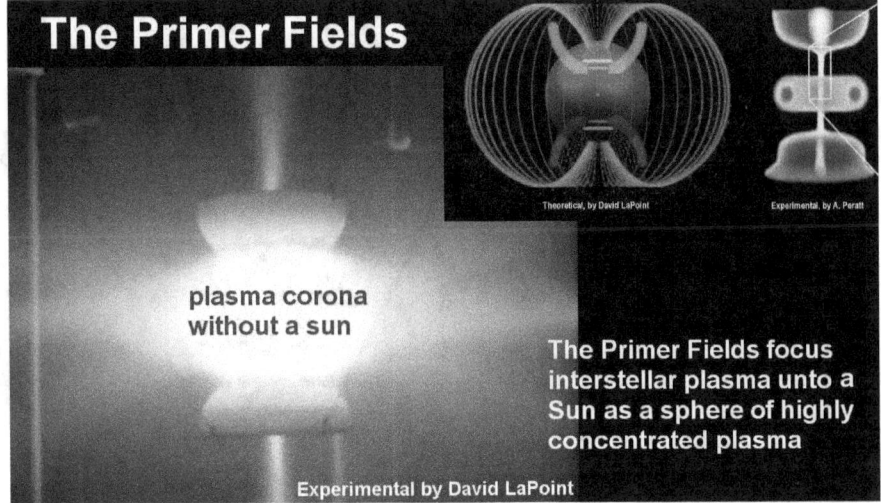

The Primer Fields

Theoretical, by David LaPoint   Experimental, by A. Peratt

plasma corona
without a sun

The Primer Fields focus
interstellar plasma unto a
Sun as a sphere of highly
concentrated plasma

Experimental by David LaPoint

David LaPoint has explored the basic structure of the Primer Fields
with a series of static experiments, in order to explore the details of
the electromagnetic geometry that has been discovered in high-
energy electric flow experiments, as shown in the example here,
produced at the Los Alamos National Laboratory, presented by
Anthony Paratt, director of experiments at the time.

A magnetic confinement dome

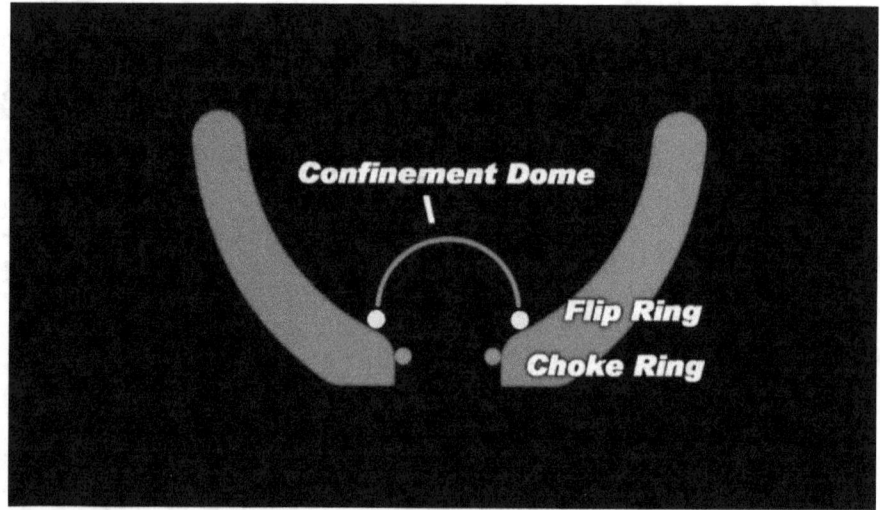

The static experiments have revealed the existence of a magnetic confinement dome.

Facilitates the concentration of plasma

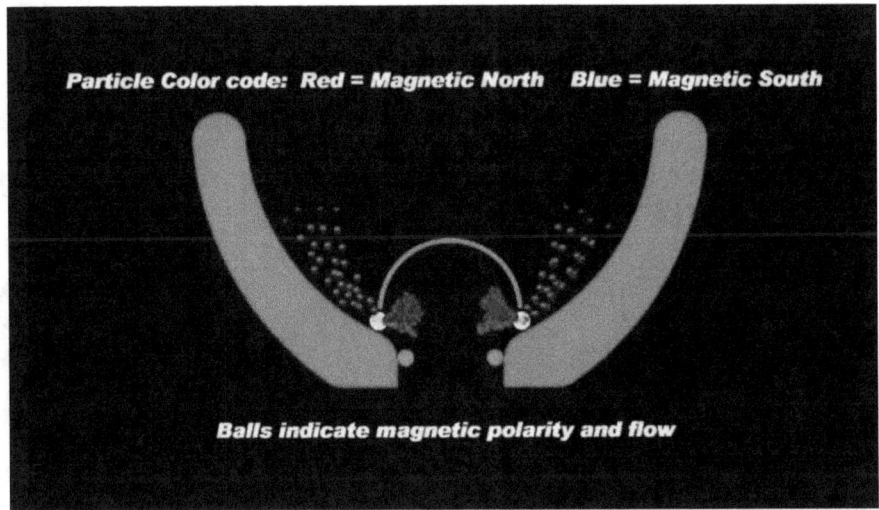

The confinement dome facilitates the concentration of plasma flowing into the system, from which the plasma becomes focused onto a sun.

David LaPoint discovered

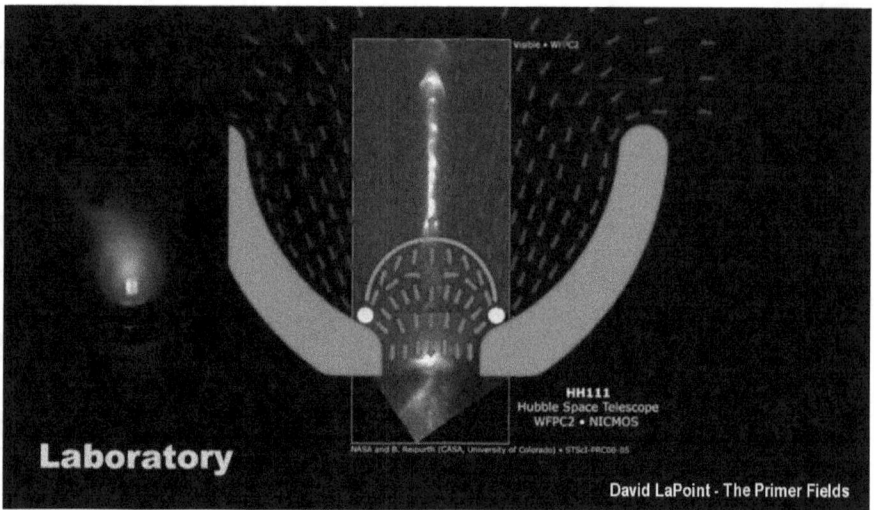

HH111
Hubble Space Telescope
WFPC2 • NICMOS

NASA and B. Reipurth (CASA, University of Colorado) • STScI-PRC98-95

Laboratory

David LaPoint - The Primer Fields

David LaPoint has also discovered that if the confined density exceeds the magnetic retention strength, some of the excess escapes through the weakest point of the confinement dome. Examples in space exist, for the resulting feature. The feature is of great value as it keeps the plasma pressure that is focused onto the Sun relatively steady, within certain limits, under normal conditions, as if the Sun was indeed a universal constant.

The Primer Fields principle

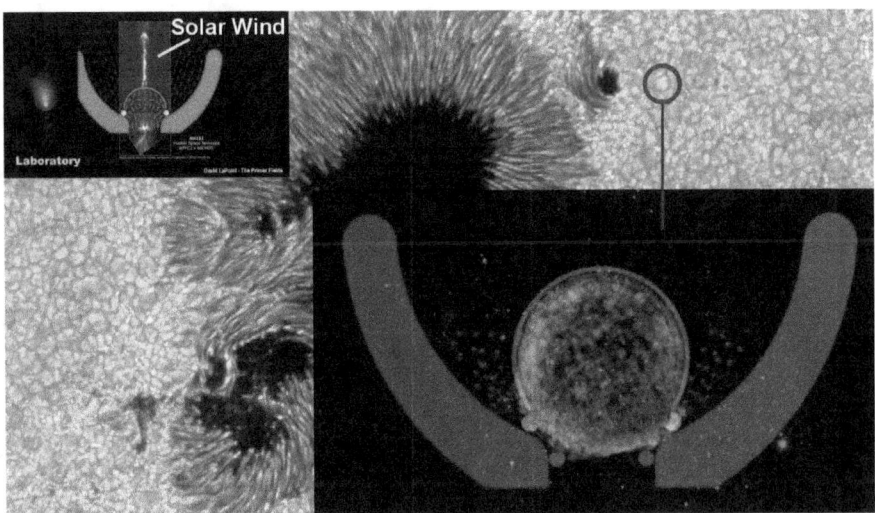

#10
The same Primer Fields principle also unfolds on the surface of the Sun, where the concentrated plasma surrounding the Sun is further concentrated to a density that enables nuclear fusion to occur, which in this case would be electrically powered cold fusion on the surface of the Sun.

Heating in the surface of the Sun

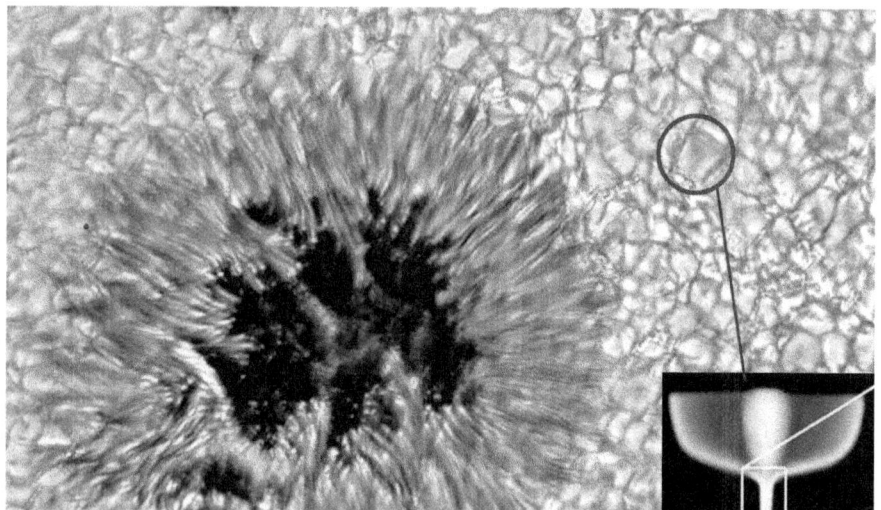

The resulting fusion process produces some mild thermal heating along the way, by the actions that cause nuclear fusion to occur. The heating in the process raises the surface of the Sun to 5,505 degrees Celsius, which falls well within the range of electric discharge temperatures.

The 5,505-degree fusion heat is extremely low, in the way that the nuclear-fusion process is theorized. The reason for this low temperature resulting from the fusion on the Sun, is that the fusion is electrically powered, rather than being thermally forced, as in the standard fusion theory, and is attempted in thermo-nuclear-fusion reactors on Earth, for which millions of degrees of temperature are required to cause any form of fusion to happen.

The Sun delivers proof

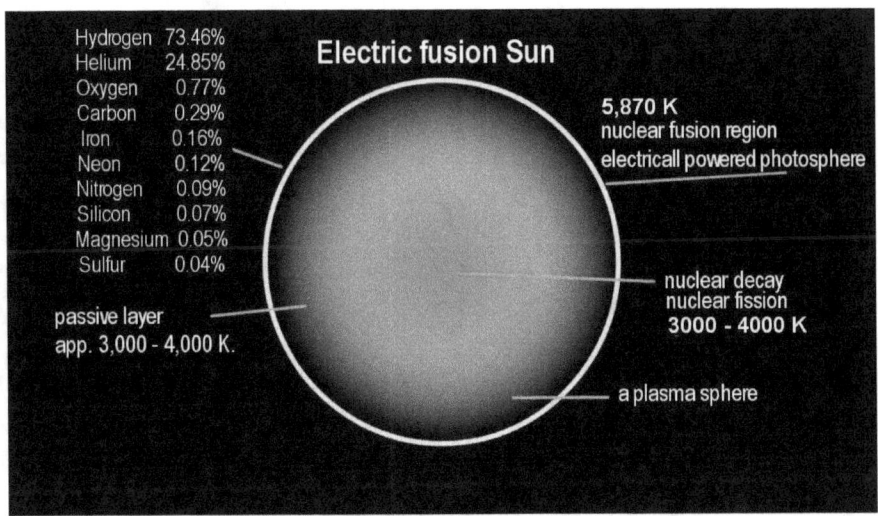

On the Sun, nuclear fusion happens naturally, and effortlessly. And above all, it operates as an atom-synthesizing fusion that creates all the atomic elements in the periodic table.

The Sun, all by itself, thereby delivers proof that we live in an electric universe.

And yes, we do have ample evidence that synthesizing nuclear fusion is occurring on the surface of the Sun.

All of these atomic elements, listed here, have been detected to be present in the atmosphere of the Sun, in the ratio shown here. If they were not synthesized on the surface of the Sun, they wouldn't be there.

The Sun takes the process no further

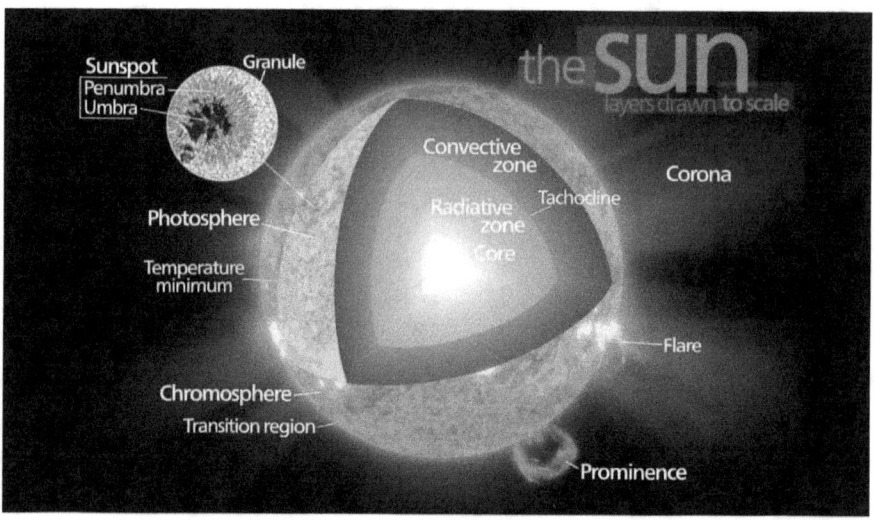

#11

The known presence of a wide range of basic atomic elements in the solar atmosphere is proof that synthesizing nuclear fusion results from the solar process, and that this is happening at the surface, not inside the Sun.

If the fusion would happen inside the Sun, the Sun would have retained the heavy elements at its core. Nor would it be possible for the Sun to produce these elements under the standard model, since the Sun is deemed to be a hydrogen fusion engine that combines hydrogen only into helium, and takes the process no further. The detected wide range of elements in the solar atmosphere simply should not exist under the standard solar model.

In the electric Sun model

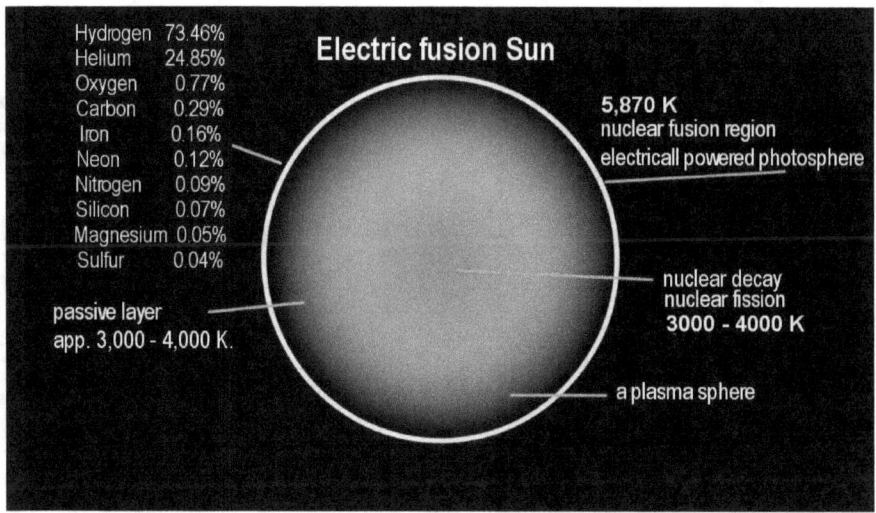

Hydrogen	73.46%
Helium	24.85%
Oxygen	0.77%
Carbon	0.29%
Iron	0.16%
Neon	0.12%
Nitrogen	0.09%
Silicon	0.07%
Magnesium	0.05%
Sulfur	0.04%

Electric fusion Sun

5,870 K
nuclear fusion region
electricall powered photosphere

passive layer
app. 3,000 - 4,000 K.

nuclear decay
nuclear fission
3000 - 4000 K

a plasma sphere

In the electric Sun model, however, where intense, electrically powered nuclear fusion takes place on the surface, the entire spectrum of elements is expected to being synthesized there, and in the ratio that is typical for the cosmic abundance of them in the solar system.

Matches elements in the cosmic abundance table

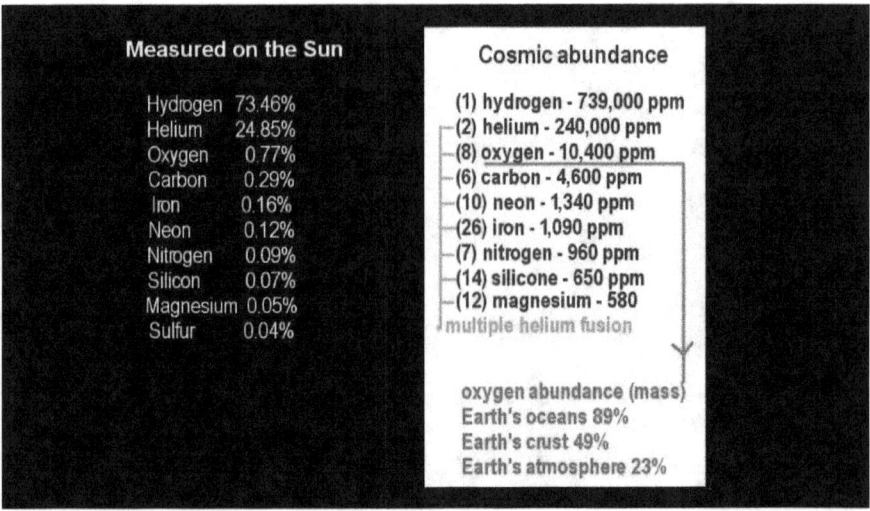

Measured on the Sun		Cosmic abundance
Hydrogen	73.46%	(1) hydrogen - 739,000 ppm
Helium	24.85%	(2) helium - 240,000 ppm
Oxygen	0.77%	(8) oxygen - 10,400 ppm
Carbon	0.29%	(6) carbon - 4,600 ppm
Iron	0.16%	(10) neon - 1,340 ppm
Neon	0.12%	(26) iron - 1,090 ppm
Nitrogen	0.09%	(7) nitrogen - 960 ppm
Silicon	0.07%	(14) silicone - 650 ppm
Magnesium	0.05%	(12) magnesium - 580
Sulfur	0.04%	multiple helium fusion

oxygen abundance (mass)
Earth's oceans 89%
Earth's crust 49%
Earth's atmosphere 23%

It is not surprising therefore that the ratio of elements detected in the solar atmosphere matches closely the known ratio for these elements in the cosmic abundance table. If these elements wouldn't be constantly synthesized at the surface of the Sun, they wouldn't exist in the Sun's atmosphere. But they do exist there, and they get carried away from there, by the solar winds. The close matching of the solar ratio with that of the abundance table suggests, that all the atomic elements that the planets of the solar system are made of, where synthesized on the surface of the Sun, rather than having been forged eons ago, during a brief instant of the mythical Big Bang explosion that supposedly created the universe.

The Sun is an electric powered star

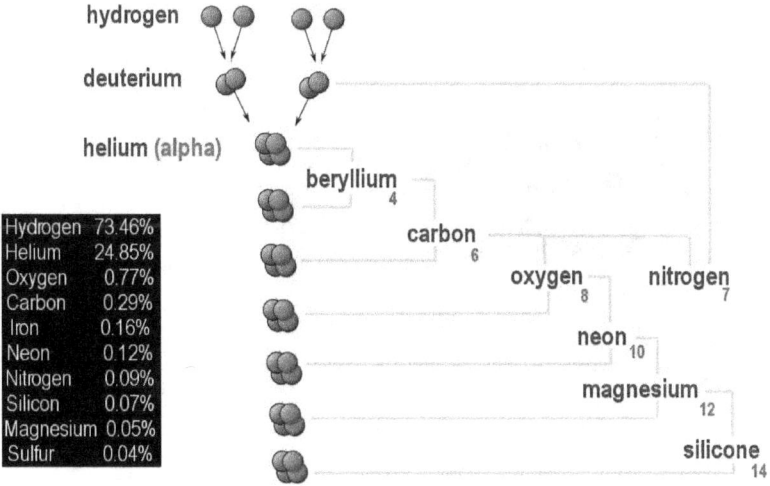

Hydrogen	73.46%
Helium	24.85%
Oxygen	0.77%
Carbon	0.29%
Iron	0.16%
Neon	0.12%
Nitrogen	0.09%
Silicon	0.07%
Magnesium	0.05%
Sulfur	0.04%

#12

The very fact, therefore, that we all exist, and that the planets exist, is evidence that the Sun is an electric powered star that over time has synthesized all the atomic elements that the planets are made of.

An internally powered, entropic Sun, does not have the capability to cause atomic synthesis.

Such a sun would therefore have no planets. Not a single solar system would exist, attached to such a Sun.

It may be for this reason that the entropic camp in astrophysical science has developed the Big Bang theory, according to which all the atoms in the universe were instantly created in a giant explosion, from which the galaxies were formed by accretion, over time, through gravitational condensation.

The Big Bang model, is dangerously entropic

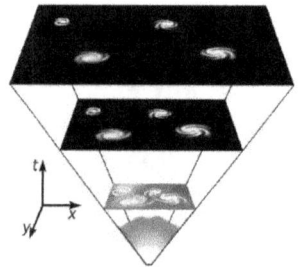

The Big Bang creation myth refuted by the electric solar fusion model

The internal-fusion theory, that is built on the Big Bang model, is dangerously entropic in that it envisions a universe that is powered by gravity-forced processes in which the universe is deemed to be consuming itself towards it eventual energy death. The acceptance of entropy as a universal principle, in which nothing is created and everything is consumed, is a dangerous fairy tale, because the tale has become subsequently accepted in economics, culture, philosophies, politics, and even in science itself. While the entropic Big Bang theory is self-evidently impossible in the real world, it does have the effect that it lulls the human spirit, even astrophysical science, to sleep. Of course, no evidence exists that the impossible has happened. The red-shift in light from distant galaxies, which is cited as evidence for the Big Bang theory, actually disproves the theory.

The red-shift phenomenon

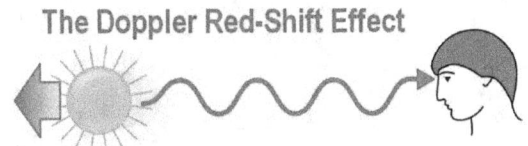

The Doppler Red-Shift Effect

light waves from a source moving away from an observer appear streched out

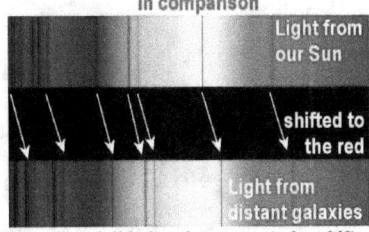

In comparison

Light from our Sun

shifted to the red

Light from distant galaxies

the entire visible band appears to be shifted

Is the phenomenon caused by a difference in speed or by a plasma-electric effect?

illustration source - wikipedia

In the entropic, Big Bang fairy dream, the Earth is once again regarded as the center of the universe, with the universe expanding around us, with growing distances everywhere, in a flow that we've got caught up in.

That's a nice tale, but not reality. The red-shift phenomenon that has been discovered in light from distant planets, which, it is said, proves that all galaxies are racing away from us into all directions, doesn't prove anything, but the opposite. It is theorized that the faster an object recedes from us, the observer, the more is the emitted light from the moving source being stretched out. It is said that the greater the red-shift is, the more has the light been stretched out, and the faster are the most distant galaxies are racing away from us.

That's the key to the Big Bang theory. But is this true?

The red-shift becomes understood

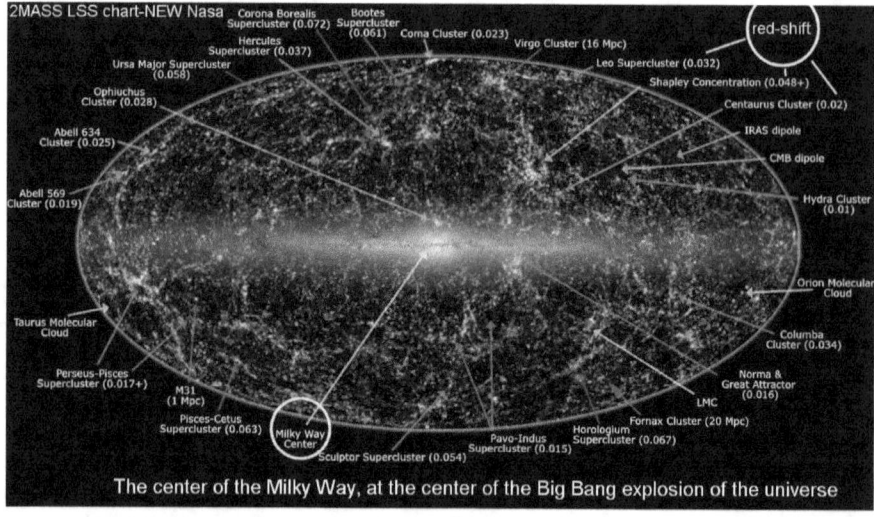

The center of the Milky Way, at the center of the Big Bang explosion of the universe

According to the Big Bang origin theory, the most distant galaxies must be receding the fastest, as if to prove that the entire universe was forged in one place at one time. That's a tale typical for bedtime stories, but in the morning the universe comes to light as a more rational place.

In the electric universe, the measured red-shift that has been observed from distant objects doesn't prove at all that the galaxies that originated the light are racing away from us.

The red-shift becomes understood as merely the result of a minuscule energy-loss during the light's propagation over long distances, through the plasma-rich intergalactic space. The energy-loss that is incurred along the way, is proportionally to the distance the light has crossed, whereby the light assumes a lower-energy state.

The photons, which are the carriers of light

Theorized image of photon

Energy Flow

image by David LaPoint - The Primer Fields

In the electric theory of the universe, the photons, which are the carriers of light, are moving points of energy that are surrounded by an electromagnetic field that give light its 'shape.'

A moving electromagnetic wave

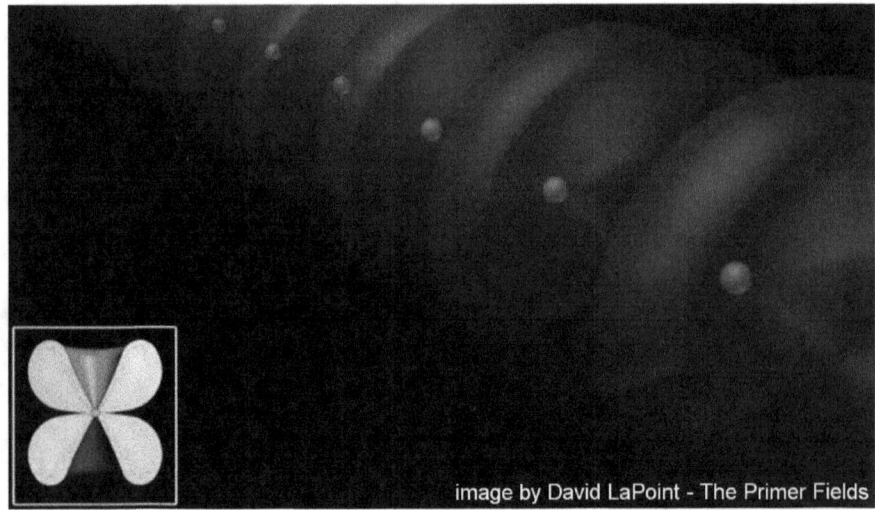

image by David LaPoint - The Primer Fields

The combination of many shapes in motion creates a moving electromagnetic wave.

The high-energy fields

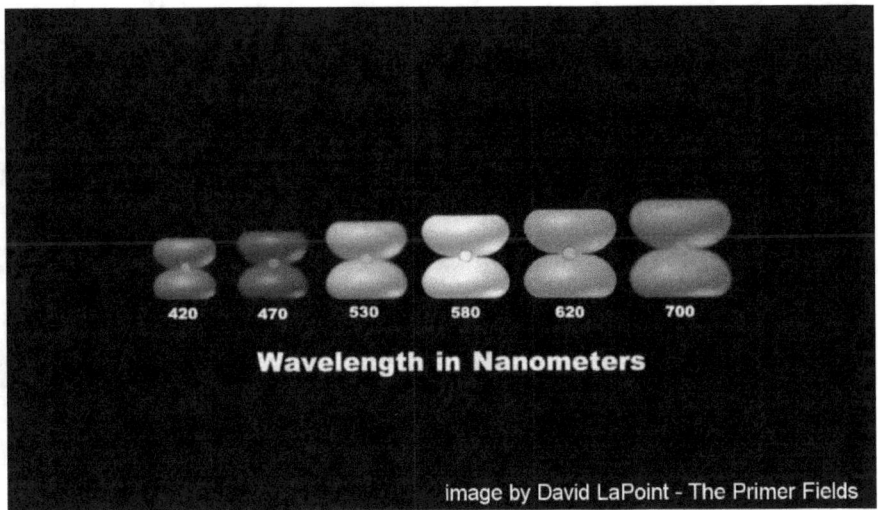

Wavelength in Nanometers

image by David LaPoint - The Primer Fields

However, not all shapes are of the same size. The differences in size determines the colour of the light. The movement of these fields, as an electromagnetic wave, determines the frequency by the size of the fields. The high-energy fields, which are more tightly packed internally, come out smaller in size.

Different wavelengths

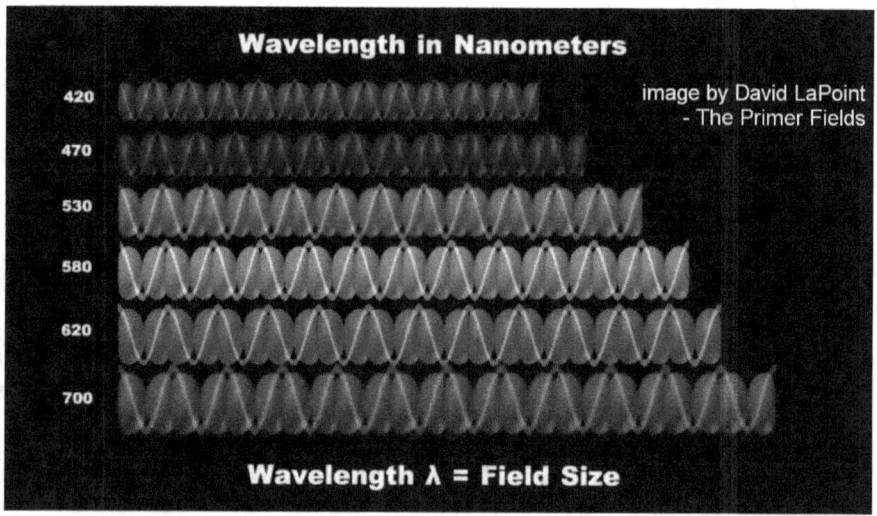

The different sizes, when strung together, create different wavelengths in their motion.

When light is propagated over long distances

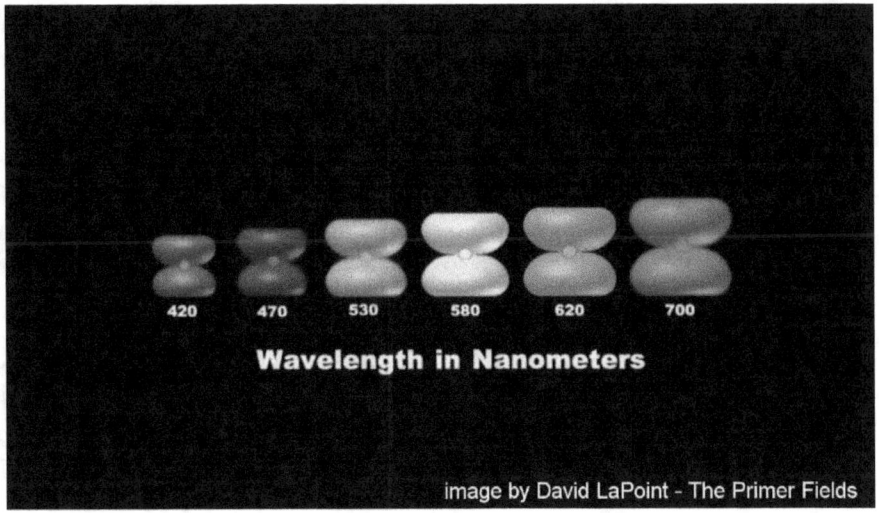

420   470   530   580   620   700

Wavelength in Nanometers

image by David LaPoint - The Primer Fields

Now, when light is propagated over long distances, such as millions and billions of light years, a gradual energy-loss is incurred along the way. The gradual energy-loss along the way results in larger electromagnetic fields, and thereby longer wavelengths. For example, the blue fields decay into the shape of the green fields, and the green fields decay into the shape of the yellow fields. In short, all the light-fields decay towards the red. The red itself decays into infrared. A shift towards the red, or red-shift for short, of all the light across the entire spectrum results from the gradual energy-loss in light propagation over long distances.

The red shift invalidates the Big Bang theory

The center of the Milky Way, at the center of the Big Bang explosion of the universe

The measured red-shift from distant objects, results from that. It is simply a measure for distance, instead of a measure of speed of separation.

The red-shift cannot be used as an absolute measure, however, as the energy-loss varies with the density of the material in cosmic space that the transmitted light encounters.

The recognition of what the red shift really represents invalidates the Big Bang theory, which is built on the red-shift phenomenon.

The recognition of what the red shift represents invalidates everything that the false theory, the Big Bang theory, has 'inspired'. And this has immense implications, because at the heart of the Big Bang theory stands the Doctrine of Entropy. The doctrine states that everything in the universe careens towards an inevitable end, as its energy stores are exhausted. Under this theory every sun burns itself out. This doctrine is applied to everything. It states that every economy careens towards its collapse; every financial system dies in debt; every nation falls into fascism; and even humanity as a whole is doomed to collapse by its increasing population that demands too many resources, so that the world must be depopulated. But the Doctrine of universal Entropy is false. The universe operates on an opposite platform, a platform of continuous development. Not a single sun burns itself out by consuming itself. The universe is dynamically self-powered, and likewise is humanity, which lives by created resources, and will do so for evermore.

Liberation from the Big Bang Doctrine of Entropy

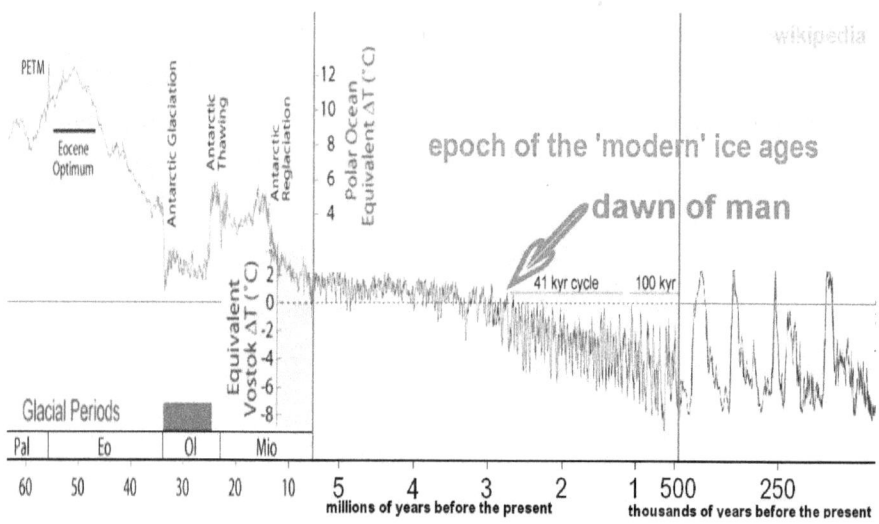

#13

The liberation from the Big Bang Doctrine of Entropy sets us free to discover how the Sun really operates, and how it can be a variable star, which the ice ages indicate it really is. Nor should the Ice Age glaciation be seen as an entropic consequence. An ice age is but a shift to a different environment that appears to be critical for human development, because humanity didn't develop itself on this planet until the modern Ice Age Epoch, the Pleistocene Epoch, began. And this too, is tightly linked to changing electric conditions in our electric universe.

Humanity is definitely anti-entropic

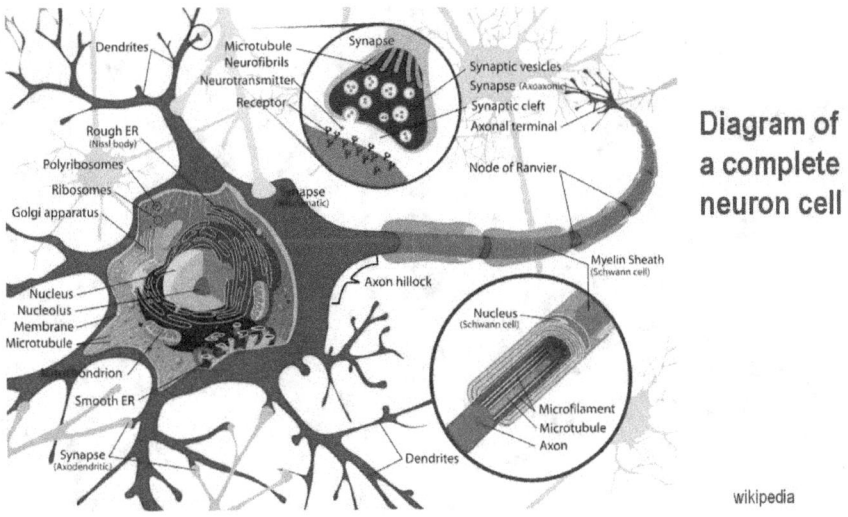

**Diagram of
a complete
neuron cell**

wikipedia

During an Ice Age, increased cosmic-ray flux is reaching the Earth and is affecting humanity. Cosmic rays are fast moving protons, which carry an electric charge. As they pass through the human body, they don't collide with anything, but they create a magnetic field that induces an electric current. The induced current may have a beneficial effect for the neurological system of the human body, which functions electrically. Humanity is definitely anti-entropic. We are in the process of developing, just as the universe continues to develop.

The term, theory, no longer applies

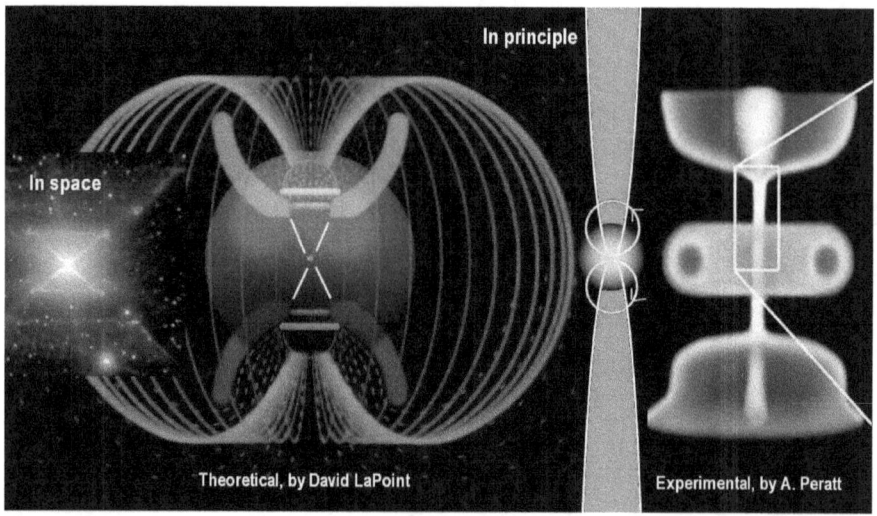

The electric universe is an endlessly developing universe. It is a construct of basic principles that apply in various contexts everywhere, from the Sun in the large context, to the photon in the smallest, with the neurological miracles of life in-between. The evidence that the universe is electrically organized on all levels is so widely apparent that the term, theory, no longer really applies, because the evidence speaks for itself.

Big Bang theory illustrates exotic excuses

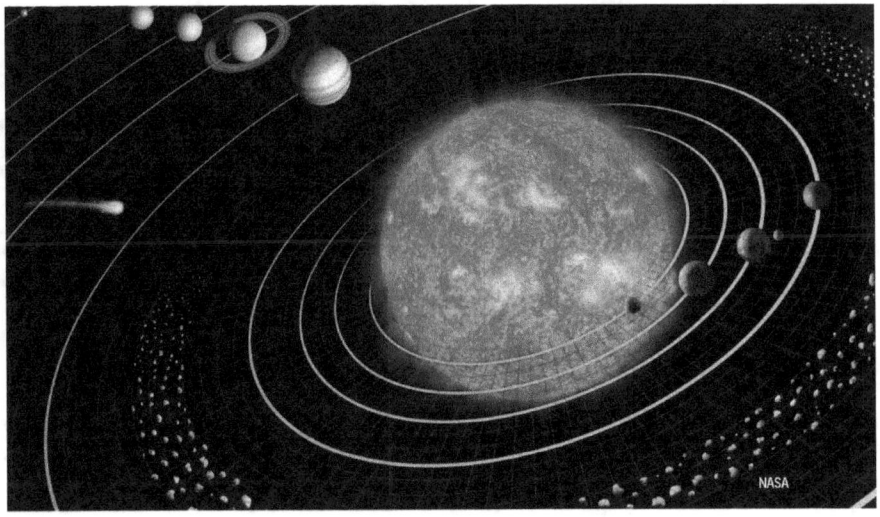

NASA

Only when we look at the universe from the standpoint of entropic philosophy, do we find ourselves boxed in with paradoxes. Those paradoxes, for which no solutions exist under the numerous false theories, typically inspire some rather exotic excuses for anything to make sense in the box. However, as the Big Bang theory illustrates, the exotic excuses ultimately don't make any sense either.

So it is, that since there exists only one type of universe, not two with opposite characteristics, we need to get out of the box and come to terms with what is real. Only one type of universe can be real. Either the Electric Universe is real - which is self-existing and self-maintaining, which all known evidence illustrates - or the entropic Big Bang dream is real, where only mass and gravity is recognized with exotic theories attached, for which no actual evidence exists. The choice shouldn't be hard.

If the Big Bang dream is real, then all of what you see and are, is an accretion of dust that was magically created in a moment of primordial explosion 13.8 billion years ago.

In this context of the empty box

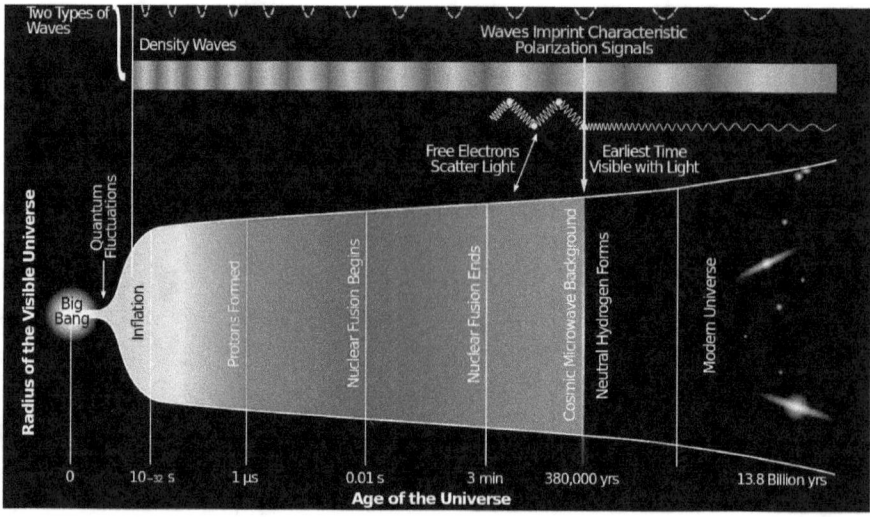

According to the Big Bang theory, all the protons, neutrons, and electrons, and so on, of the entire universe, were formed in one place in the first one-millionth of a second of the explosion, which then was driven apart by a gravity shockwave that is said to be still expanding.

In this context of the empty box, the electric force of the universe is omitted as a factor, though it is 39 orders of magnitude stronger than gravity and its effects. By this omission, which makes the box rather empty and useless, the Big Bang theory renders itself as but an empty dream, and likewise everything that is built on this dream.

Sun in a still-ongoing dynamic process

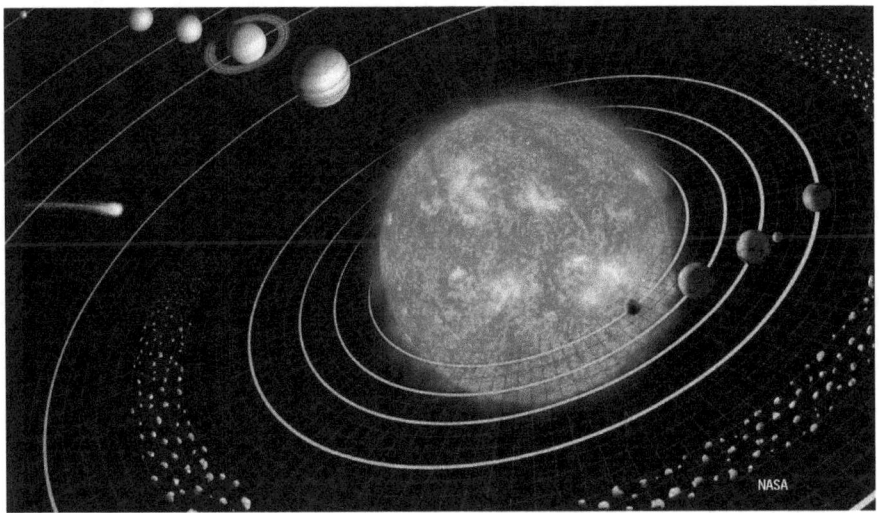

In the real universe the creative process is continuously unfolding. The physical evidence that we see in our solar system tells us that all atomic elements that make up the planets were electrically synthesized on the surface of the Sun in a still-ongoing dynamic process that we have physically measured the result of.

The alpha fusion chain

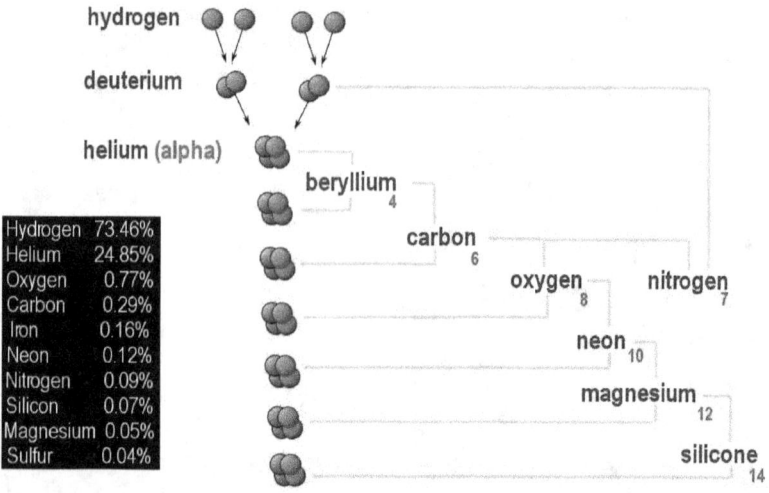

Hydrogen	73.46%
Helium	24.85%
Oxygen	0.77%
Carbon	0.29%
Iron	0.16%
Neon	0.12%
Nitrogen	0.09%
Silicon	0.07%
Magnesium	0.05%
Sulfur	0.04%

The alpha fusion chain that is illustrated here in simplified form, illustrates the basic process that has formed the planets over long periods, all forged from basic plasma, powered by in intense electric interaction. The entire solar system, as we see it today, was likely produced rather quickly over a few million years while the Sun itself was formed from concentrated plasma streams near the center of the galaxy in the highest plasma-dense region. The creative process for the solar system is illustrated in the distribution of atomic elements across the different planets, understood as flowing from the Sun.

The distribution pattern proves

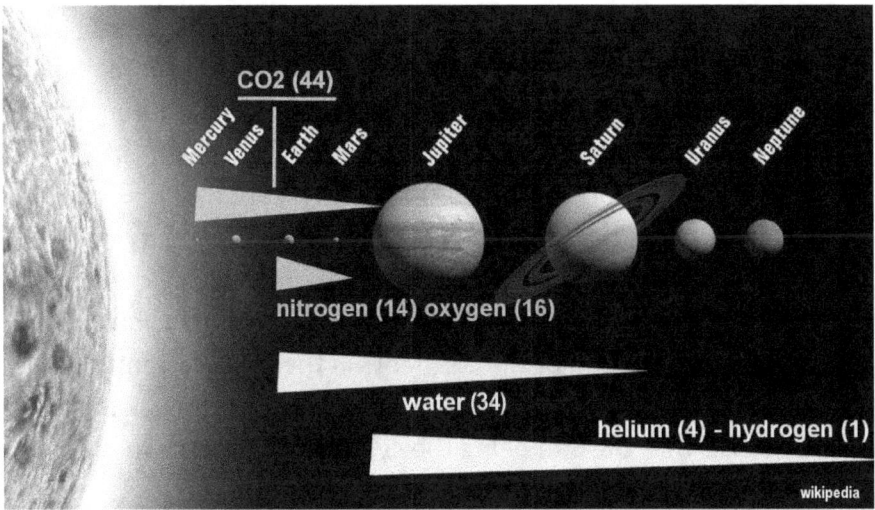

CO2 (44)

Mercury Venus Earth Mars Jupiter Saturn Uranus Neptune

nitrogen (14) oxygen (16)

water (34)

helium (4) - hydrogen (1)

wikipedia

#14

The distribution pattern proves that we live in an electrically powered universe in which a sun is the synthesizing engine that produces, in surface nuclear fusion, all the atomic elements that the planets are made of. The distribution is necessarily uneven. The heavier elements are evidently the first to fall out, from the solar winds that carry them, by gravitational attraction, and so on.

This means that the heavy, rocky planets with metallic cores and the heaviest gases in the atmosphere, are located nearest to the Sun, while the planets of the lighter elements are formed farther away. It is evidently for this reason that the atmosphere of Venus is made up almost entirely of CO2, a heavy gas with the combined atomic weight of 44.

In this context the Earth would have captured most of the atomic nitrogen and oxygen, which are somewhat lighter in atomic weight. Since most of the oxygen produced on the Sun would have been combined along the way, with hydrogen from the Sun, to form water, the fallout of water would have to be strongly apparent on the Earth and the planets nearby extending as far as Saturn that has rings and moons made of accumulated water ice.

The fallout of the much 'lighter' hydrogen and helium atoms, evidently begins most strongly at a greater distance from the Sun. Here the giant gas planets are formed, with Jupiter capturing most of the helium and hydrogen atoms, and Saturn, Uranus, and Neptune becoming progressively smaller thereafter. This distribution ratio all by itself disproves the Big Bang theory and proves that we live in an electric universe with an electric Sun that synthesizes atomic elements at its surface by electric fusion.

The unique distribution of the elements in the solar system, that has been discovered with extensive exploration missions, only makes sense with the Sun being the synthesizing source for the distributed elements.

Ratio of atomic elements

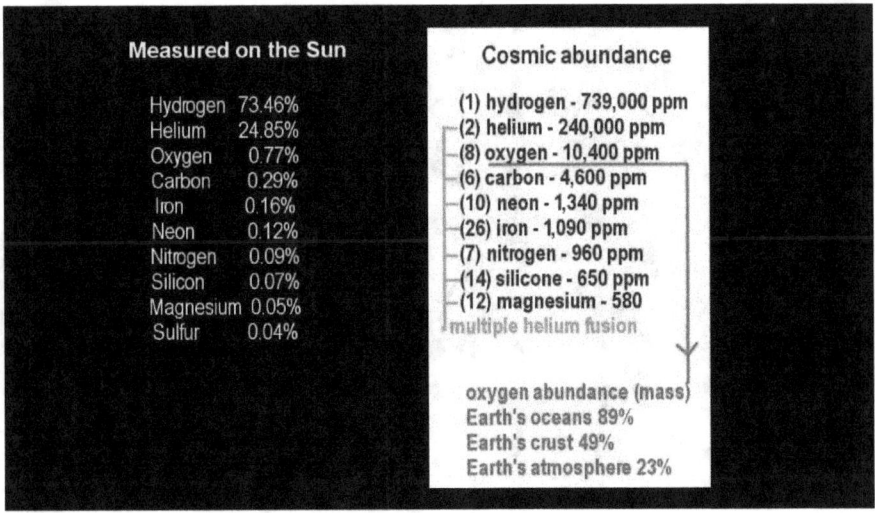

Measured on the Sun

Hydrogen	73.46%
Helium	24.85%
Oxygen	0.77%
Carbon	0.29%
Iron	0.16%
Neon	0.12%
Nitrogen	0.09%
Silicon	0.07%
Magnesium	0.05%
Sulfur	0.04%

Cosmic abundance

(1) hydrogen - 739,000 ppm
(2) helium - 240,000 ppm
(8) oxygen - 10,400 ppm
(6) carbon - 4,600 ppm
(10) neon - 1,340 ppm
(26) iron - 1,090 ppm
(7) nitrogen - 960 ppm
(14) silicone - 650 ppm
(12) magnesium - 580
multiple helium fusion

oxygen abundance (mass)
Earth's oceans 89%
Earth's crust 49%
Earth's atmosphere 23%

Measurements also tell us that the ratio of atomic elements existing in the solar 'atmosphere,' matches closely the theorized ratio published as the global abundance table.

Atomic elements flowing away from the sun

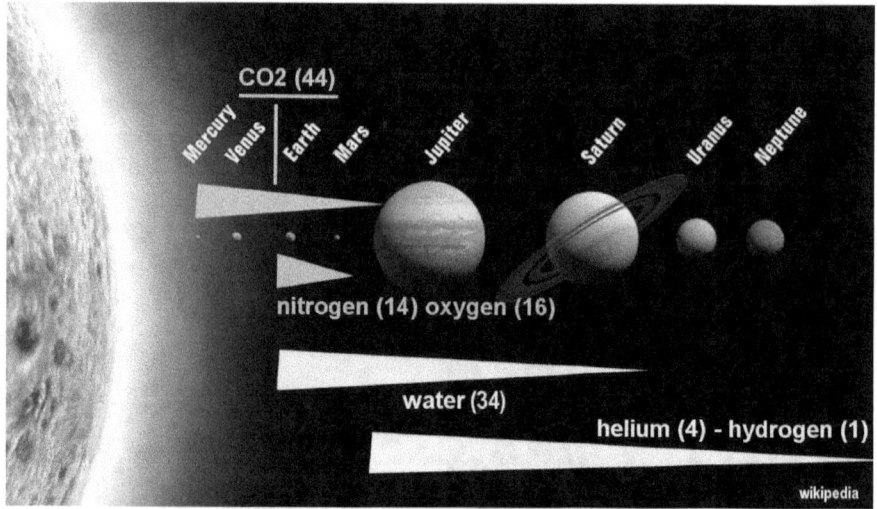

The distribution pattern of the atomic elements flowing away from the sun, which is thereby evident as the synthesizing source, adds one more item of proof that we live in an electrically powered universe.

Of course, the universe doesn't end at our doorstep, nor do its principles and their manifestations end here.

Distribution of synthesized atomic elements

M45 CC BY 3.0 wikipedia - Permission from Filip Lolic

#15

The distribution of synthesized atomic elements is evidently not confined to the solar system exclusively, especially in the case when larger stars produce larger amounts of synthesized atomic elements that are electrical neutral and are thereby free flowing, and flowing off into space far past their respective solar systems. Some of the larger stars form clouds of gas and dust around them, and flowing away from them.

In entropic science

In the case of extremely large stars, and entire clusters of stars, the resulting clouds of synthesized atomic elements take on gigantic proportions. It is widely believed in entropic science that these clouds are remnants of exploding stars. But if one looks closer, they all have super-active giant stars still at their center, or associated with the massive clouds of synthesized gases and heavy elements, termed dust.

Enormously large clouds of synthesized elements

Tarantula Nebula in the Large Magellanic Cloud with the R136 cluster,
R136a1/2/3 are visible as the barely resolved knot at bottom right.
The brightest star just to the left of the cluster core is R136c, and
another extremely massive WN5h star.
ESO/P. Crowther/C.J. Evans

Enormously large clouds of synthesized elements have accumulated
over time near clusters of super-giant stars, as in the case shown
here. Some of the giant clouds have remained largely stable. Other
clouds have been seen expanding with the typical speeds of flowing
solar winds.

Synthesized giant clouds also reveal

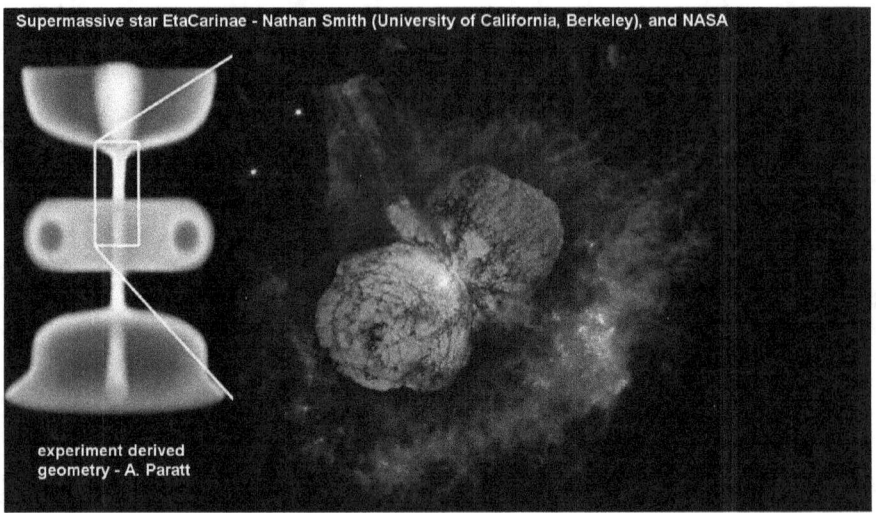

Supermassive star EtaCarinae - Nathan Smith (University of California, Berkeley), and NASA

experiment derived
geometry - A. Paratt

#16

In some cases the synthesized giant clouds also reveal the tell-tale geometry of the plasma flow structures that form them, as the geometry has been discovered in plasma-flow experiments in laboratories, such as the one shown here derived from experiments at the Los Alamos National Laboratory.

The double-bowl geometry of the Primer Fields

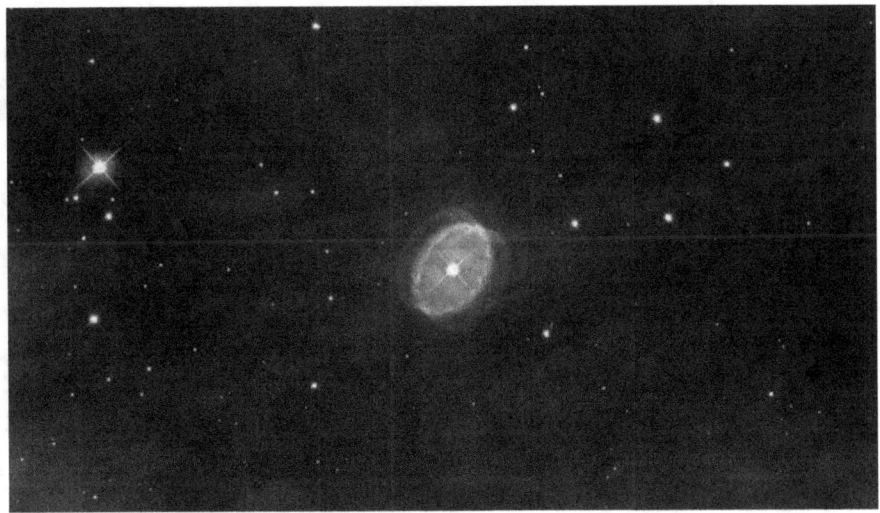

In the case of some smaller giant stars, the double-bowl geometry of the Primer Fields that focus plasma onto a sun, remains still visible. The two large scale examples illustrate visibly the active plasma geometry of the Primer Fields that focus plasma unto a sun, including our Sun.

Plasma inflow geometry has been physically verified

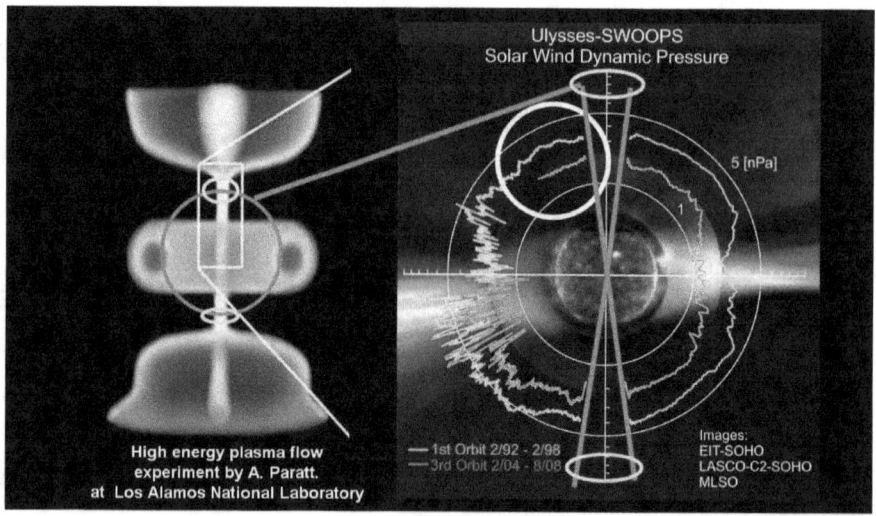

High energy plasma flow experiment by A. Paratt. at Los Alamos National Laboratory

Ulysses-SWOOPS Solar Wind Dynamic Pressure

5 [nPa]

— 1st Orbit 2/92 - 2/98
— 3rd Orbit 2/04 - 8/08

Images:
EIT-SOHO
LASCO-C2-SOHO
MLSO

#17

In the case of our solar system, the plasma currents are too weak to be visible. However, the plasma inflow geometry has been physically verified by NASA's Ulysses spacecraft in the form of the measured interruptions of the solar wind in the areas where the plasma inflow to the Sun would be located. What we see measured, delivers physical proof that our local star, the Sun, is unmistakably, an electric star.

The phenomenon of the noctilucent clouds

Noctilucent clouds over Kuresoo bog, Viljandimaa, Estonia, app. 75-85 Km high - wikipedia

#18

The evidence that synthesizing fusion is still happening on the surface of the Sun, is also seen in the phenomenon of the noctilucent clouds, which are clouds of water ice in the upper stratosphere up to 80 kilometers above the surface. Water vapor, which is heavy, isn't lifted up as high as to the very edge of the atmosphere. However, in the electric solar system where atomic elements and molecules flow from the Sun, pervading interstellar space, the existence of water ice at the edge of the Earth's atmosphere, is extremely natural.

The rings of Saturn

Saturn's rings
93% water ice

#19
The rings of Saturn are similarly composed of water ice. They are items of proof that we live in an electric solar system. The water for the rings was evidently accumulated over long periods as deposits from the Sun.

Saturn's moon, Dione

The high water concentration in the Saturn system is further evident in Saturn's moons. The entire moon, Dione, is made up almost entirely of water ice.

Saturn proves with its moons

Saturn's moon Tethys

1060 Km diameter
composed mostly of water ice

The same high concentration of water ice is found on the moon, Tethye, which is another one of the moons of Saturn. These giant moons are made of ice from water that originated at the Sun in the form of hydrogen and oxygen, which evidently combined into water along the way. Saturn proves with its moons that we live in an electric universe, under an electric Sun, as no other cause can create the kind of phenomena we see.

Looking through the umbra of the sunspots

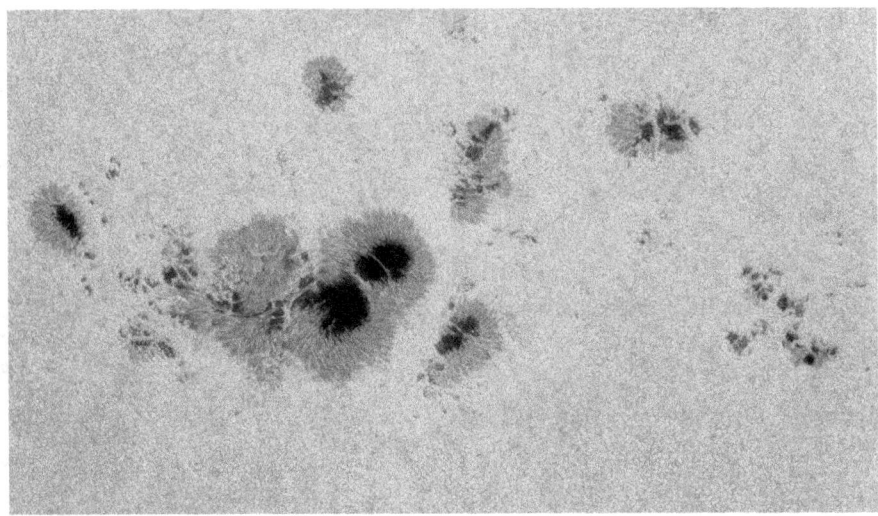

#20

Of course it is also fairly evident by simply looking at the Sun, that the Sun is not internally heated. By looking through the umbra of the sunspots, at what lies below the shiny surface, we see a dark and cold interior. What we see isn't a paradox in the electric sun model. It is expected there. It is a paradox only in the internally heated fusion model, which is contradicted thereby.

The Sun a plasma sphere

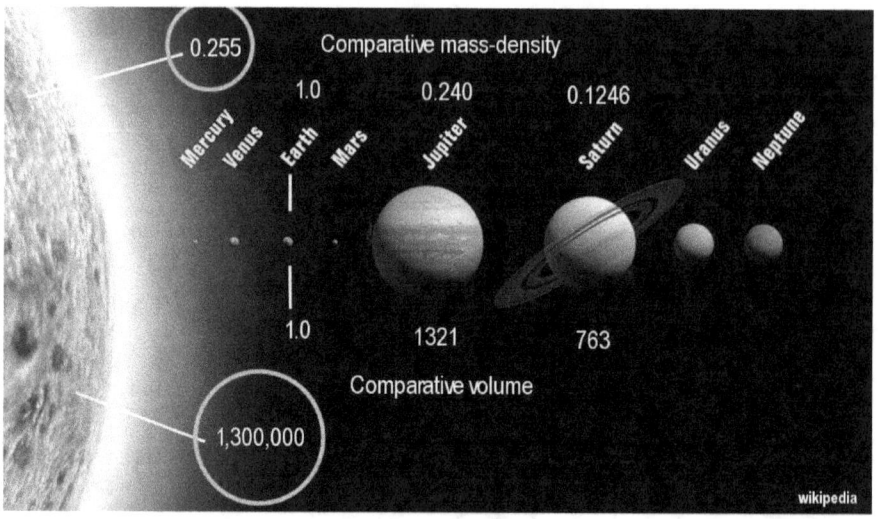

Comparative mass-density

0.255
1.0   0.240   0.1246

Mercury Venus Earth Mars Jupiter Saturn Uranus Neptune

1.0   1321   763

Comparative volume

1,300,000

wikipedia

#21

The internally heated fusion model also contradicts itself by its very premise. For the internally forced hydrogen fusion to occur in the Sun, the Sun needs to be a sphere of atomic hydrogen gas, but which is not physically possible. The enormous gravity of a gas sphere the size of the Sun, would crush the atoms that it would be made of. And even if it was possible for such a gigantic gas sphere to exist, the resulting, extremely compressed gas, would make the Sun up to a thousand times heavier that it actually is.

The extremely light weight, in mass-density comparison with Jupiter and Saturn, that the Sun is known to have, renders the Sun a plasma sphere that is diffused by the force of electric-repulsion, a force that is 39 orders of magnitude stronger than gravity, which would enable a sun of its size to exist, as well as any other imaginable size, even super-massive stars.

Stars with a mass 150 times greater

Tarantula Nebula in the Large Magellanic Cloud with the R136 cluster,
R136a1/2/3 are visible as the barely resolved knot at bottom right.
The brightest star just to the left of the cluster core is R136c, and
another extremely massive WN5h star.
ESO/P. Crowther/C.J. Evans

#22

For example, Stars with a mass 150 times greater than the mass of
the Sun, have been located, like the stars in the R136 Cluster. It
takes a giant leap of faith to believe that these gigantic spheres of
atomic gases can exist without its atoms being crushed by the
ensuing gravity. The largest of these, R136a1 is believed to be 300
times more massive than the Sun.

The very existence of such giant stars proves that the stellar
universe is electrically powered as no other reasonable option exists
for these giants to be formed.

A simple size comparison

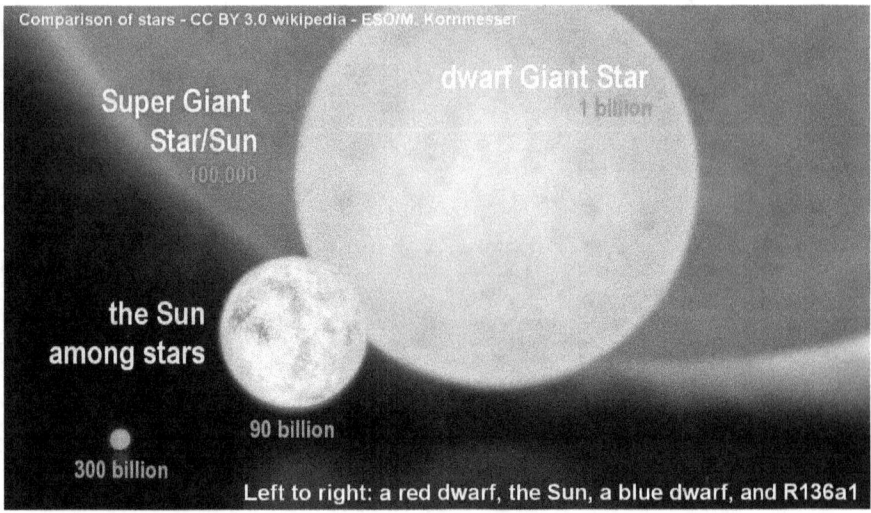

Super Giant Star/Sun
100.000

dwarf Giant Star
1 billion

the Sun among stars

90 billion

300 billion

Left to right: a red dwarf, the Sun, a blue dwarf, and R136a1

Just look at a simple size comparison, like the one shown here. Using a combination of instruments on the European Southern Observatory's Very Large Telescope, astronomers have discovered the most massive stars to date, some weighing more than 300 times the mass of the Sun, or twice as much as the previously accepted limit of 150 solar masses. This artist's impression presented here shows the relative sizes that are known for stars, from the smallest 'red dwarfs', weighing in at about one tenth the mass of the Sun, through the low mass "yellow dwarfs" such as the Sun is regarded to be, ranging up to the massive "blue dwarf" stars weighing eight times more than the Sun, and from there to the 300 solar mass super giants, like the one named R136a1. Plasma spheres of this supergiant size are inherently possible, while gas spheres are not, except in the imagination. The very fact that this wide range of stars exists, renders every sun to be an electrically powered star, based on the same universal principle. It is a well proven fact that electric phenomena are possible on almost any scale. Some stellar principles can be replicated even in the laboratory.

Concepts of the entropic star cycle

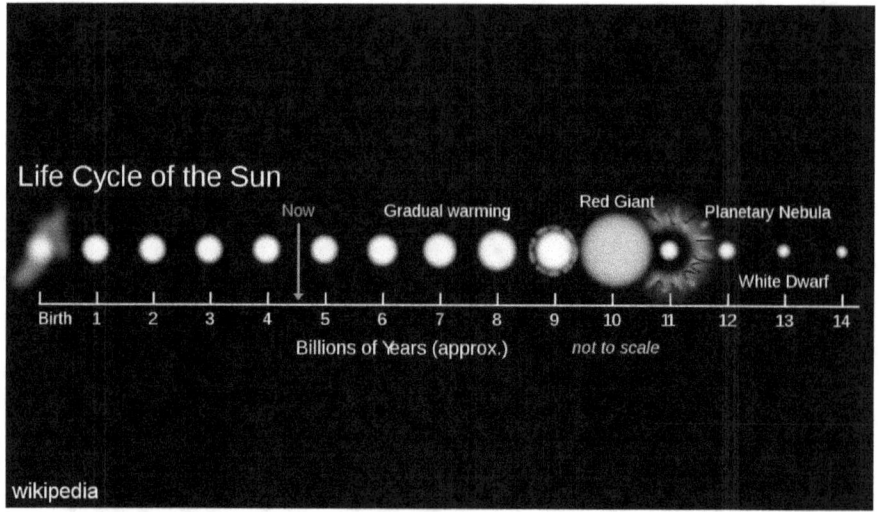

The stellar physics concepts of the entropic star cycles, are concepts that are boxed into recognizing nothing but gravity-forced reactions. The box is largely empty, as it takes 99.999% of the universe, which exists in plasma form, out of the theoretical considerations. What comes out of the largely empty box is a distorted view of the universe. The distortion may have a political motive standing behind it. It is a well-established fact that every system of empire, regardless of its form, forces entropic conditions onto the economies the system of empire controls and exploits. The imperial system champions the concept of the entropy as a closed in system of thinking in which everything is winding down to an inevitable depletion of its very substance. This may be the reason why the Big Bang theory was hastily promoted at the time when the electric universe theory was beginning to unfold, to counter-act it. The successful promotion, however, does not mean that the politically motivated distortion of science is correct.

The wide scene of Ice Age historic evidence

The wide scene of Ice Age historic evidence speaks to us of a solar system that is powered by an electric Sun that is inherently a variable sun.

Since all the measured Ice Age evidence contradicts the accepted theory about the nature of the Sun, the time has come to explore what is really happening. We need a theory that supports the known evidence. We need it, in order that we stop lying to ourselves. This necessarily includes the way we regard the solar system, because the Sun and the solar system are an integrated whole. Since the long accepted theory of the internal-fusion sun is wrong, let's scrap it as a construct of delusions, no matter how challenging this surrendering may seem. The delusions reflect the notion that only gravity rules in the cosmic universe.

Cosmic universe is ruled by a much greater force

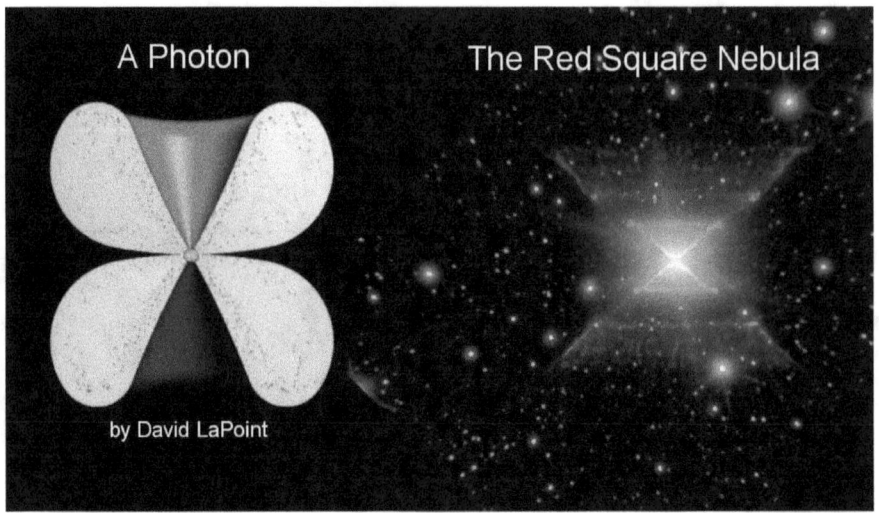

A Photon   The Red Square Nebula

by David LaPoint

#23

The evidence tells us that the cosmic universe is ruled by a much greater force, which is the electric force, that is as powerfully evident on the cosmic scale as it is on the atomic scale. Not a single atom would exist without the electric force being expressed by its principles.

The ecliptic phenomenon is expected

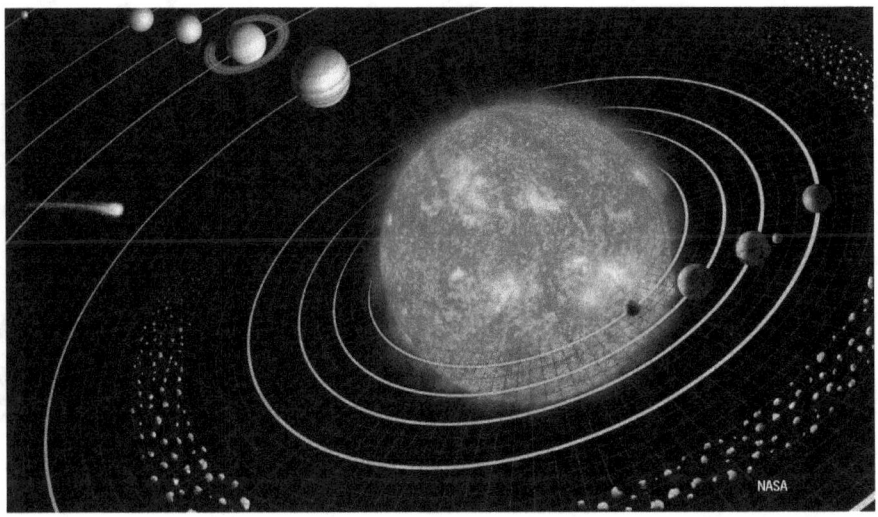

#24

In a similar manner, a solar system would not exist if its operating principles did not exist, which are electric principles. No gravity-mass principle exists, or is possible, that would, for example, dictate that the planets in a solar system must orbit orderly together in a 'thin' ecliptic plane.

In the electric universe, of course, the ecliptic phenomenon is expected. It is actually rather simple there.

Saturn's rings are made of water ice

Saturn's rings
93% water ice

Let us look at a simple example, such as the case of Saturn that has a strongly visible system of rings that is perfectly aligned with the planet's equator. Saturn's rings are made primarily of water ice. Why would these rings be aligned with the planet's equator? The rings extend from 6,630 km to 120,700 km from the planet, but are extremely thin comparatively, at an average thickness of only 20 meters. But what keeps this sheet so perfectly flat? The answer is located in electrodynamics.

The ecliptic phenomenon

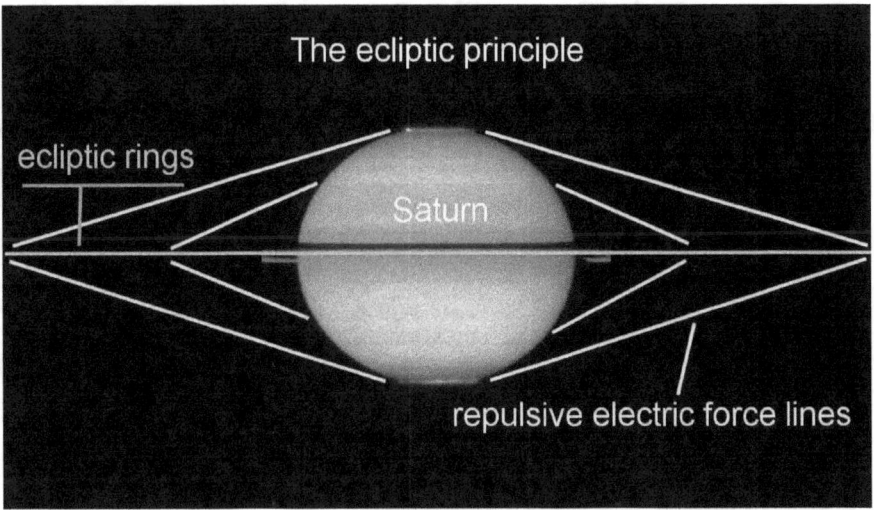

The phenomenon has been experimentally verified in high-power electric-flow experiments. The ecliptic phenomenon is not possible on any other platform than the electromagnetic platform. With the planets being bombarded with the solar wind that consists to a large degree of protons, the planets and all its moons and whatever interstellar material it attracts, becomes proton-rich. It thereby gains an electric charge. This means that a repelling force acts against the material of the rings from above and from below in such a manner that the repelling forces balance each other. And since the planet is spinning, the movement of the protons in the plant surface creates a moving electric field that becomes electromagnetically coupled into the orbiting ring system and planets.

The thin disk of the galactic plane

Andromeda Galaxy

Galaxy NGC 5866

The same principle also creates the thin disk of the galactic plane. No other known principle would be capable of squishing an entire galaxy of 400 billion stars into a thin flat disk. Our galaxy, the Milky Way Galaxy is roughly 100,000 light years wide and 1,000 light years thick, a 100 to one ratio. The missing element that would cause the flattening of the galaxy into a thin disk, has only recently been discovered.

Two plasma confinement domes of the Primer Fields

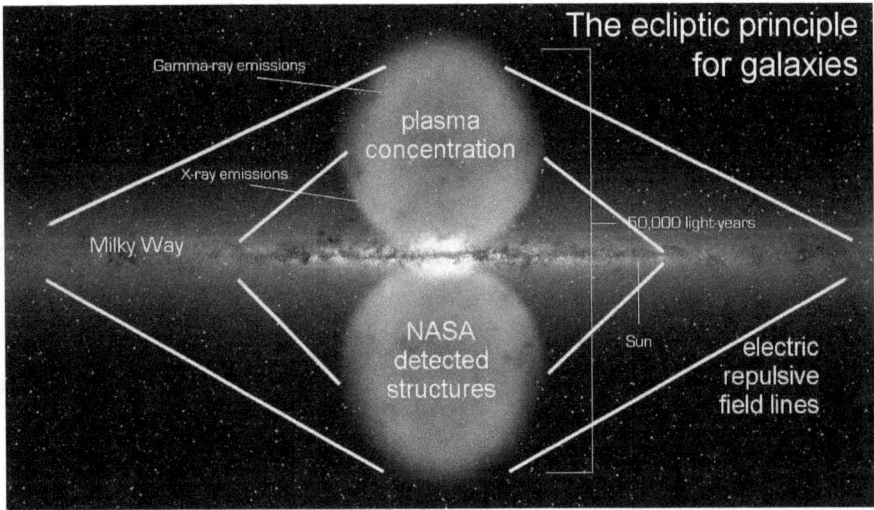

#25

The recently discovered element consists of two plasma confinement domes of the Primer Fields for the galaxy, which stand like giant towers above and below the galaxy. These structures span 25,000 light years in each direction. These giant fields of concentrated intergalactic plasma, are extremely proton-rich. The principle of the electric repulsion, acting from above and below, appears to force the stars into a vast ecliptic disk, as one of the principles involved. The principle appears to be identical to the principle that creates the ecliptic rings around planets. The electric ecliptic principle appears to be so universally manifest that it applies almost everywhere in cosmic space where the requisite conditions exist. Obviously, the same principle also applies to the planetary ecliptic of the solar system where similar electric characteristics of the Primer Fields apply as on the galactic level, which focus a large sphere of plasma around the Sun.

Exceedingly puzzling in entropic cosmology

Andromeda

Saturn

All of this means that the entire universe is electromagnetically organized. Every major feature that we see in the cosmos, especially those that appear exceedingly puzzling in entropic cosmology, is evidently electromagnetically produced, without exception, illustrating that the universe is anti-entropic in nature. A part of the proof of the principle is found in the fact that the electric ecliptic principle is reflected equally in the very large scale on the galactic level, as it is on the very small scale, as in the case of the planets' ring systems.

The moons of the planets

NASA

Gossamer Rings

Halo

Main Ring

Amalthea

Adrastea

Metis

Jupiter
its rings and planets

Thebe

#26

All of the big gas planets in our solar system have ring structures, not only Saturn. Most are less visible, as they are of lesser density. However, regardless of the density of the rings, in each case, the ecliptic disk is aligned with the planets' equator, and in each case the moons of the planets orbit within this tightly confined ecliptic plane.

The rings of Jupiter are fainter

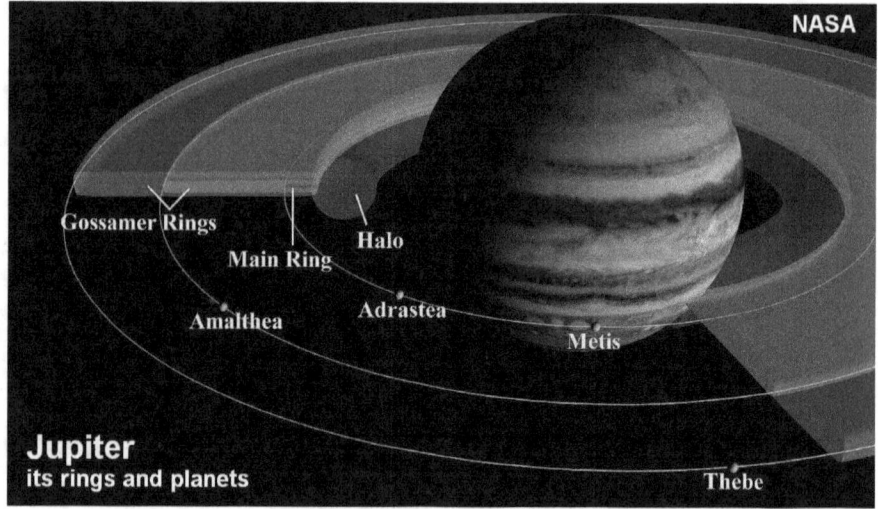

On Jupiter, a number of moons are directly a part of the ring structure. The rings of Jupiter are fainter than those of Saturn as they are made of fine dust, but they do exist.

The rings of Uranus

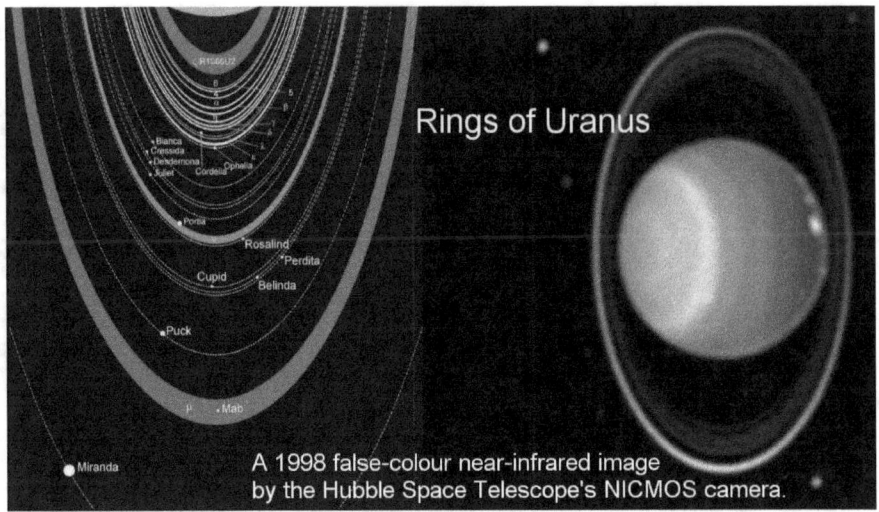

Rings of Uranus

A 1998 false-colour near-infrared image
by the Hubble Space Telescope's NICMOS camera.

#27
The universal ecliptic principle is also evident in the rings of Uranus.
Even though the planet has an extremely backwards tilted spin axis,
the plane of the rings tilts with the spin axis and remains equatorial.
Most of the moons of Uranus orbit within the ring system itself.

The rings of Neptune are still fainter

Rings of Neptune
This is a 591 second exposure by the Voyager 2 wide angle camera

NASA

The rings of Neptune are still fainter. But they do exist. They have been photographed by satellites. Evidence also exists that suggests that a similar electromagnetic coupling as the one that drives the ring system, also powers the movements of the big storms in the gas planets' atmosphere. Neptune holds the record in this category. It features the strongest storms in the solar system, even while it is the farthest away from the Sun

Neptune is the most distant planet

Neptune is the most distant planet of the solar system. It is 30 times as far from the Sun than is the Earth, and it is nearly 4 times larger. It has 14 moons, the largest is named Triton.

The coldest object found in the solar system

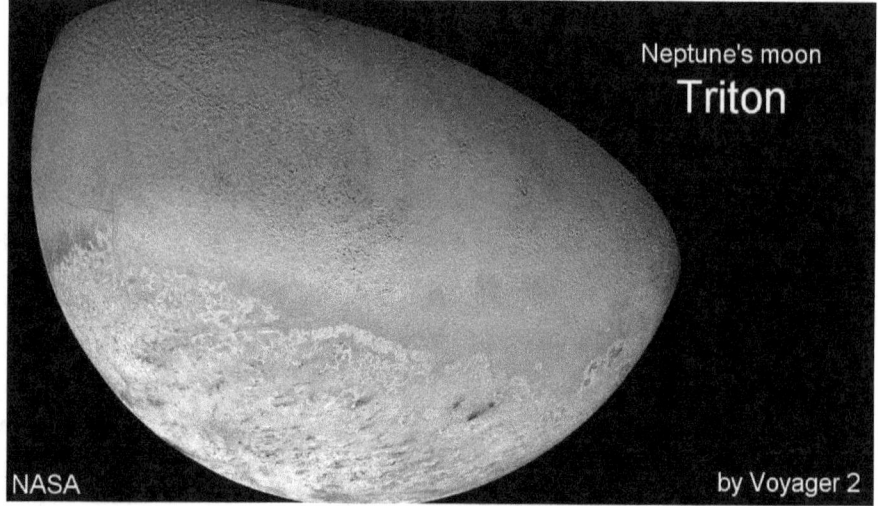

Neptune's moon
Triton

NASA

by Voyager 2

This moon is the coldest object found in the solar system with temperatures in the range of minus 235 degrees Celsius - a mere 38 degrees above absolute zero. Ironically, its cold parent-planet features winds measured up to 2,100 kilometers an hour. How is this possible? The Neptune system is so far distant from the Sun that it gets almost no solar heat.

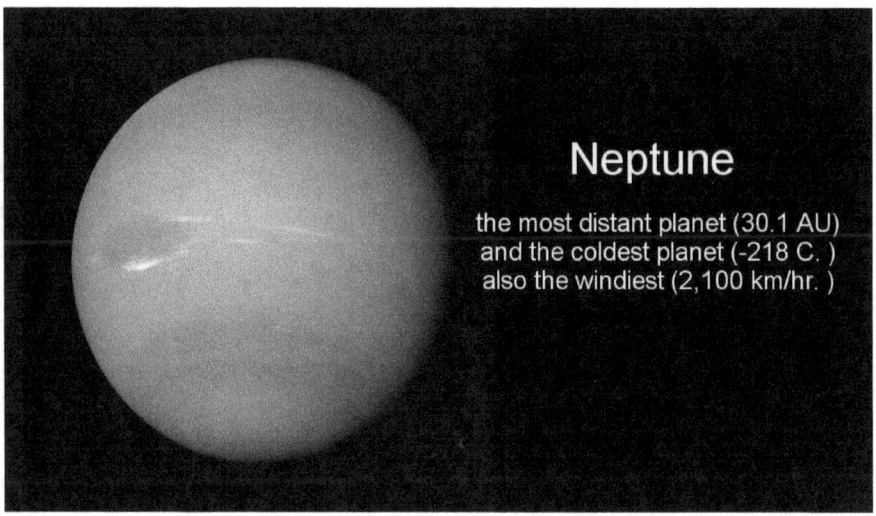

Neptune

the most distant planet (30.1 AU)
and the coldest planet (-218 C.)
also the windiest (2,100 km/hr.)

#28

Storms shouldn't be happening at all on Neptune. What would power the storms then, when Neptune's upper atmosphere is stone cold, at minus 218 degrees Celsius? What gigantic power source, powers the violent, supersonic air movement on the planet at up to 2,100 kilometers an hour? Storms of this gigantic magnitude can only happen in this icy landscape when they are of electric origin, rather than being a thermal phenomenon.

Oddly, the key to the answer to what powers the storms lies in the orbital distance of Neptune from the Sun.

By the principle of magneto hydrodynamics

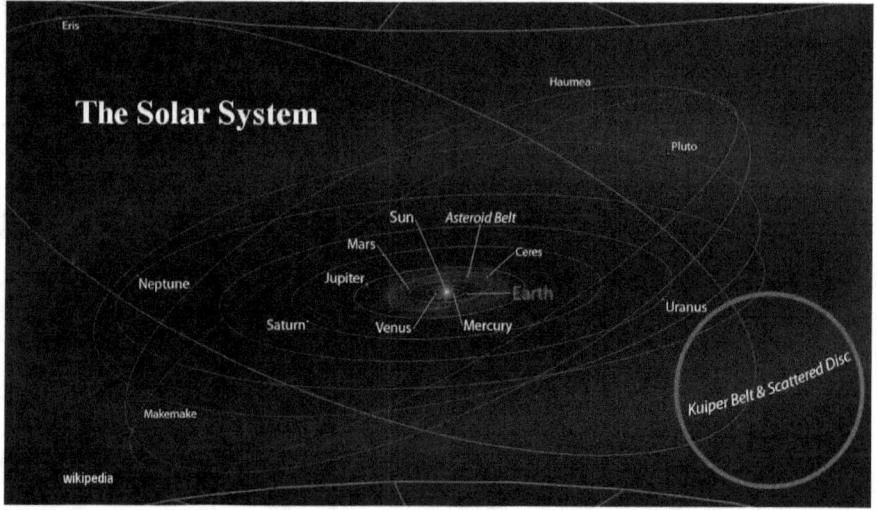

At the extreme distance of Neptune's orbit, the solar winds have likely slowed enough so that Neptune gets a larger dose of its protons than any of the other planets. The larger dose renders the planet more electrically active. Depending on the solar wind fluctuations, the winds on Neptune can be driven to exceptionally high speeds by the principle of magneto hydrodynamics, causing atmospheric movement far in excess of the surface rotation of the planet.

Something makes Saturn special

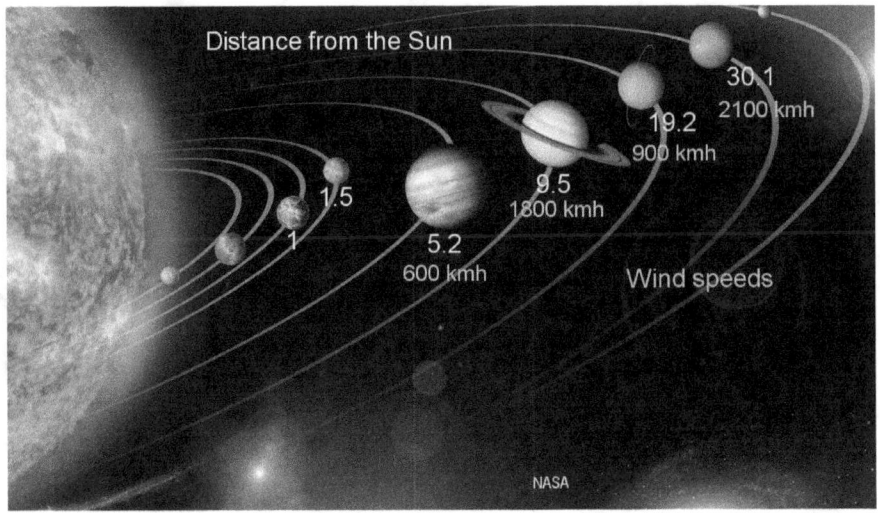

Distance from the Sun

1

1.5

5.2
600 kmh

9.5
1800 kmh

19.2
900 kmh

30.1
2100 kmh

Wind speeds

NASA

#29

A principle of relationships comes to light when one compares the storm speeds for the planets with the planets' distance. The measured storm speed increases with distance, which would correspond with increased proton density on the planets. Only Saturn doesn't follow the pattern. Something makes Saturn special.

Saturn's exceptional storms

Saturn Storms

view from NASA's Cassini spacecraft. Image credit: NASA/JPL-Caltech/SSI

#30

Saturn's exceptional storms, for its closer position, appears to be due to Saturn being located at a distance where the external magnetic fields, from the Primer Fields, are extremely weak. This renders Saturn the only planet whose effective magnetic field coincides almost perfectly with its dynamo magnetic field. Its effective deflection is a mere 7-tenth of a degree, even while the spin axis of the planet is strongly tilted at 29 degrees. Under such conditions, with almost no external field interfering, Saturn appears to be able to attract larger volumes of solar wind, which gives the planet a richer proton density across its surface. In this case, the storm front is moving against the rotation of the planet, exceeding its fast rotational speed.

Saturn's exceptionally strong polar aurora

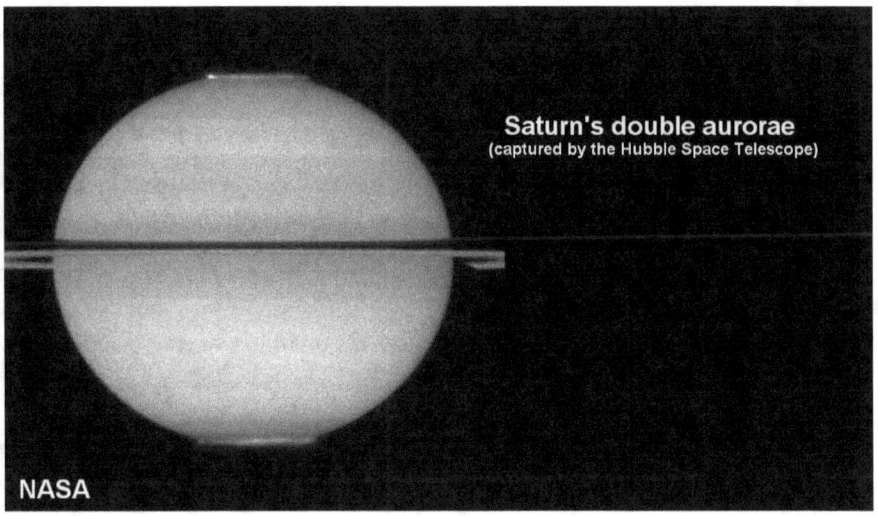

Saturn's double aurorae
(captured by the Hubble Space Telescope)

NASA

#31

The resulting larger, proton inflow, from the solar wind, is also likely the cause for Saturn's exceptionally strong polar aurora.

Aurora phenomenon on Jupiter

Jupiter aurora

NASA - HST

#32

The aurora phenomenon that is visible on Saturn, on both poles, is similarly visible on Jupiter, only to a lesser degree, comparatively.

Jupiter's storms are nevertheless gigantic

Jupiter
its rings and planets

#33

This weaker plasma influx on Jupiter is likely the cause for its 'weaker' storms, which are nevertheless strong as we see them encircling the planet. Though the move with less than supersonic speeds, Jupiter's storms are nevertheless gigantic storms by Earth standards with speeds, in the 360 kilometre per hour range, reaching upward to 600 kilometers per hour.

Jupiter is a cold planet

Infrared of Jupiter by
ESO's Very Large Telescope.

Jupiter MAD - wikipedia CC BY 3.0

The storms on Jupiter, of course, are likewise not caused by solar thermal influx. Jupiter is a cold planet with extremely weak solar energy input that heats the planet to a maximum temperature, at the cloud surface, of minus 108 degrees Celsius. Nevertheless, in the ultra violet image that is shown here, the polar regions, which should be colder, show up as being 'warmer' due to the planet's attraction of electric energy from nearby space, by its large gravity and strong magnetic field. The light band at the equator indicates that strong electric interaction is taking place there, right at the equator, where the electric system interacts with the ring system. This is also the region where the strongest cloud movement on Jupiter is occurring.

Aurora on Earth are relatively weak

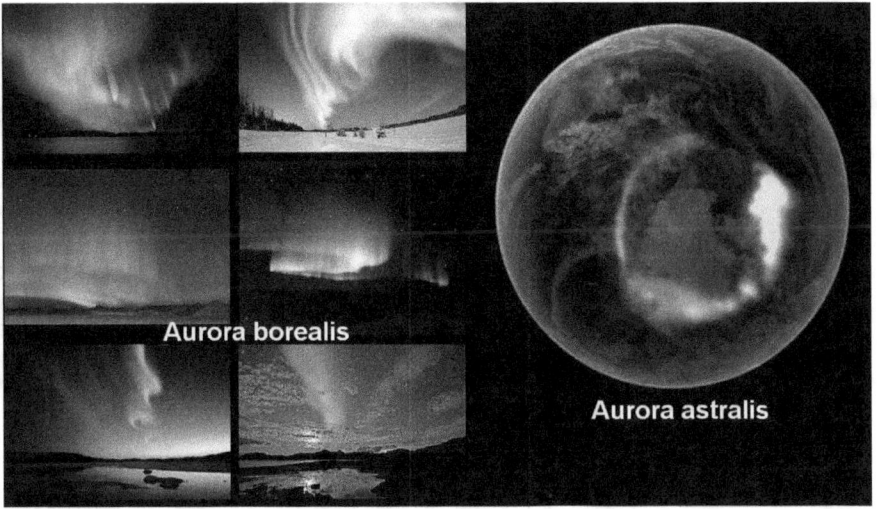

Aurora borealis

Aurora astralis

#34

The aurora on Earth are relatively weak, except in times of coronal mass ejections that are extraordinary events, which fall into a different category than the steady background aurora. This means that electric events that we see on Earth, follow the same pattern that we see expressed in principle throughout the solar system. Nor is the universality of electric evidence surprising, as the entire solar system itself is the product of electric actions on the Sun that synthesized all the atomic elements for the planets of the solar system in the first place. We can't get away from the fact that the creative, electric force of action is the organizing force in the universe, including in our galaxy, and in our solar system within it.

Perpetual motion machines are not possible

#35

It is self-evident by looking at the various types of electric evidence that we would not have a planetary system orbiting our Sun, if the orbits of the planets were not electromagnetically assisted in their orbital motions. In physical mechanics, perpetual motion machines are not possible. Every mechanistic system is inherently entropic in nature, meaning that it winds itself down by energy depletion like a wind-up toy spends the energy invested into it. But this is not how a solar system operates. It does not diminish.

Our technological satellites that are orbiting the Earth, are all entropic systems. The satellites are doomed to fall back to earth as their kinetic energy is drained away by their encountering plasma in space, and atomic elements flowing in the solar winds. Every satellite is doomed thereby, unless its orbit is actively assisted by re-boosting. The same physical entropy applies also to the planets. Every planet, regardless of its size, ploughs through fields of plasma interwoven with atoms and molecules which drain its kinetic energy. The loss of energy would de-orbit the planets over time if the energy drain was not compensated by an offsetting active force. The amazing stability of the solar system, that has been spanning billions of years, speaks to us of powerful electromagnetic assistance being constantly applied., which provides for its stability. This means that the entire solar system is actively powered. Nothing happens passively here. Everything happens actively in the solar system for as long as the solar system itself remains powered. Here is where our vulnerability begins, because our Sun is a mediocre small star among the stars in the galaxy. Its size is typical for a weak plasma environment. This environment is presently diminishing towards a point at which the Sun will become inactive completely, as it apparently has through 85% of the last million years of the current Ice Age Epoch.

But before we can explore our weakening Sun system, lets look at the big star systems for comparison.

In the Cat's Eye nebula

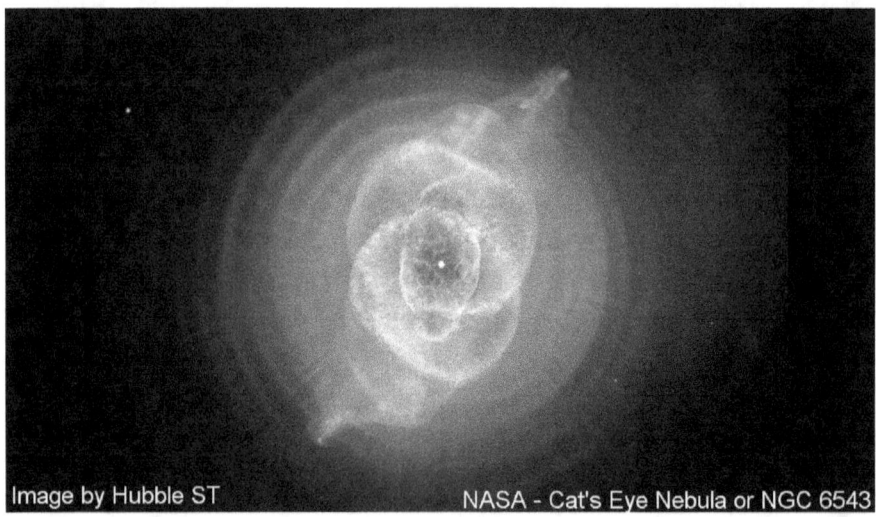

Image by Hubble ST   NASA - Cat's Eye Nebula or NGC 6543

#36

Images like these are often deemed to be the result of an exploding star, or a partial explosion that 'liberates' vast quantities of its atomic gases and materials. But is this true? In the electric universe were planets are formed from atomic materials that have been synthesized at the surface of a sun, for which ample evidence exists, one doesn't require exploding stars as a source for the glowing atomic materials that nebulas are made of. One only requires a large star and large plasma flows being focused on it. In the Cat's Eye nebula, about 3300 light-years away from the Earth, the central star there is believed to be smaller in radius than our Sun, but is 10,000 times as luminous than the Sun is.

This 'miracle', all by itself, is not possible under the entropic-sun theory. A sun operating at a temperature of approximately 80,000 degrees, in comparison with the 5,500 degrees of our Sun, while this sun has half the mass, is not even imaginable under the gravity-forced fusion theory. The 10,000 times more active star in the nebula is proof that we live in an electric universe. It is self-evident too, that the heating of the surrounding material, which is heated to 9,000 degrees, and remains heated with out diminishment, is not residual heat that would remain for but a moment, but is actively heated by inflowing plasma interacting with atomic materials in the solar winds. Even the solar winds itself, moving at speeds in excess of 1,900 kilometers per second, would not provide this tremendous active heating. It is estimated that the solar winds in this nebula, carry atomic material away from its sun at a rate of 20 trillion tons per second. The energy and mass for this enormous rate of nuclear-fusion synthesis is derived from cosmic sources, cosmic plasma, flowing into the nebula, and to its sun.

Some light nuclear fusion synthesis

Cat's Eye Halo - CC BY 3.0 - Wikipedia - Nordic Optical Telescope and Romano Corradi

The synthesized material that spreads continuously outward across a wide area until it drifts out of the range of the plasma currents to illumine it, is not derived from a small star shedding its skin, but is electrically formed. Some light nuclear fusion synthesis may also be happening in the plasma streams themselves, by the plasma that powers their luminance. This, however, is merely a theory.

Researcher David LaPoint regards a nebula

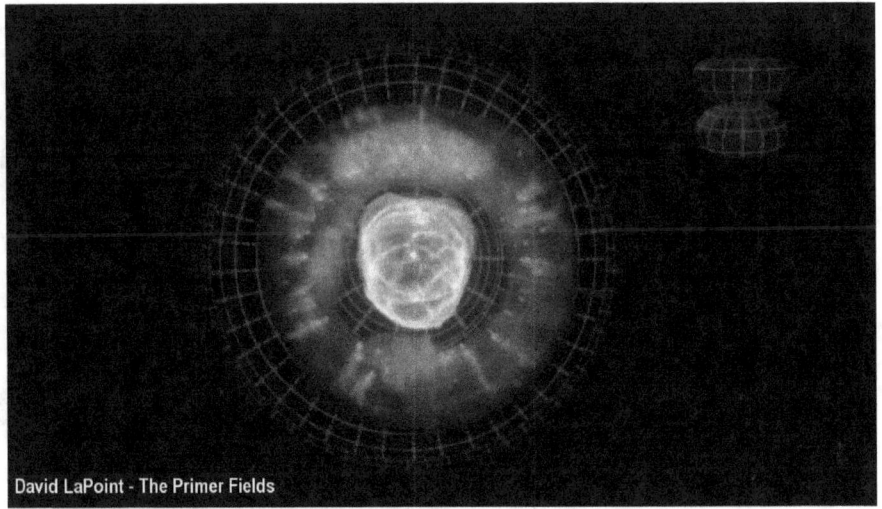

David LaPoint - The Primer Fields

The plasma researcher David LaPoint regards a nebula on the scale of the Cat's Eye nebula, to be the focal point of correspondingly large structures of Primer Fields, as he has illustrated here in principle. This principle evidently applies to nebulas of all sizes, all the way up to the largest galaxies.

The famous Crab nebula

The Crab Nebula

NASA

#37

A somewhat larger nebula, than the Cat's Eye nebula, though of essentially the same type, is the famous Crab nebula. The nebula may have started with the convergence of large plasma streams that gave rise to the birth of a powerful star that is still ongoing. The nebula is sometimes referred to as a cosmic "generator" that is producing energy at the rate of 100,000 suns combined.

The entire galaxy is in motion.

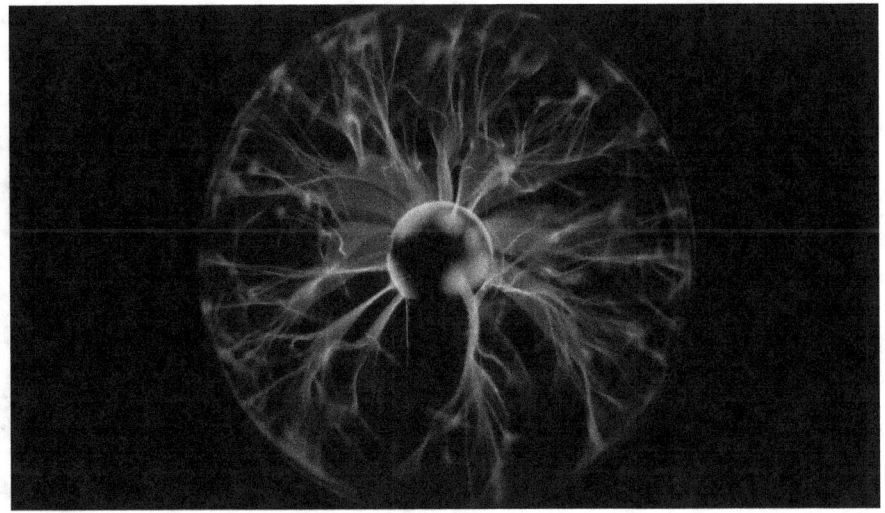

Plasma streams in space are constantly in motion as this plasma sphere as a toy illustrates. The entire galaxy is in motion.

The plasma streams in the galaxies

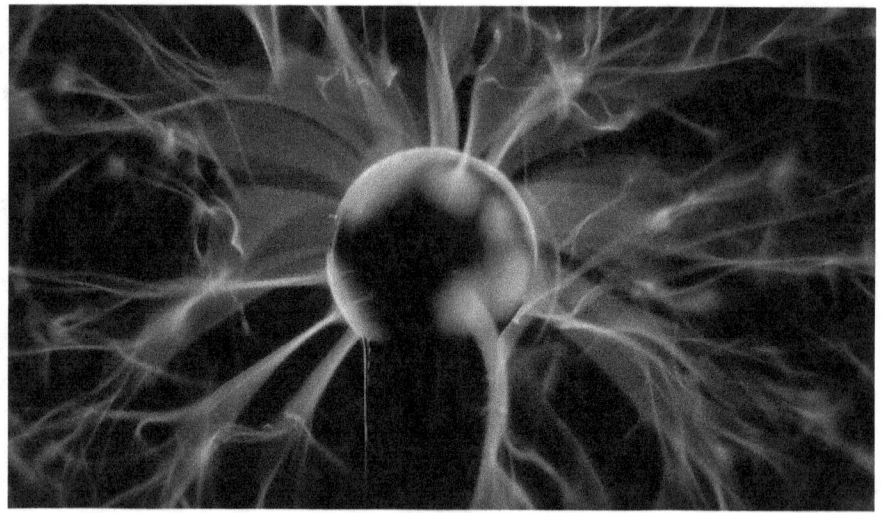

The plasma streams in the galaxies are always twisting, sometimes breaking, sometimes reconnecting.

Stars cannot explode

When large streams converge, large events happen, even explosive events that are termed supernova events. These events are not powered by exploding stars. Since stars are inherently spheres of plasma, as gas spheres of the size of stars cannot exist, stars cannot explode, because plasma spheres do not explode. Supernova events are therefore merely large plasma-convergence events.

A puff of smoke

A puff of smoke

ESA/Hubble & NASA
Acknowledgement: Claude Cornen   SNR 0519 - located over 150 000 light-years from Earth

#38

Some of the big flashy plasma events are actually successful in creating stars. The Cat's Eye nebula is evidently an example. Some lesser events, however, are clearly not extensive enough to meet the minimal threshold requirement, and end up as but a puff of smoke.

Smoke from the supernova 1987A

Supernova 1987a

09/1994

NASA - Oct 2011

#39

A somewhat larger puff of smoke resulted from the supernova 1987A. But it too, fell short of what is needed to create a star. Still, the case is interesting as evidence that we live in an electric universe.

A burst of neutrinos from the so-called supernova reached the Earth on February 23, 1987, followed by a bright light three hours later. It is believed that the neutrino emission resulted from the core collapse inside a star. It is theorized that the emission of light was delayed, as it occurs only after the resulting shock wave reaches the stellar surface that rips the planet apart. However, it is far more likely that the neutrino emissions occurred simultaneous with the transition of light, because in a plasma event they would have occurred together, so that the measured time gap reflects a minute difference in propagation speeds.

Light is slowed by denser media

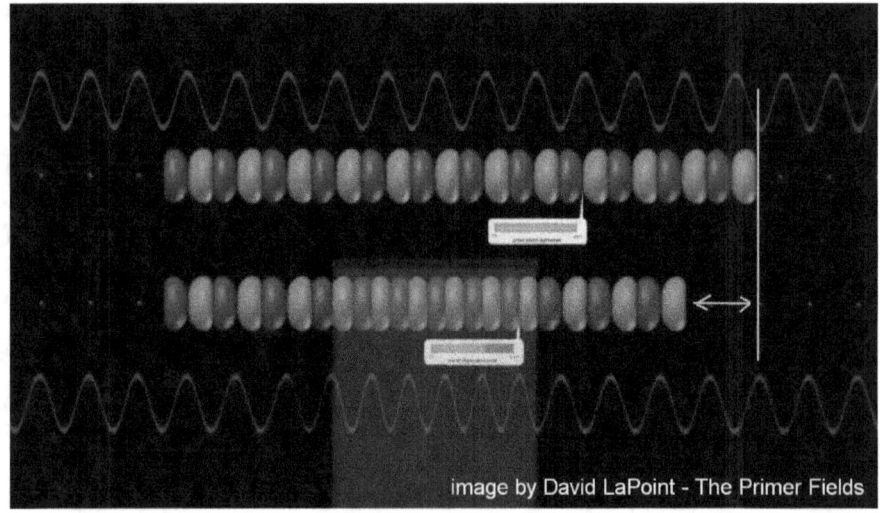

image by David LaPoint - The Primer Fields

Light is slowed by denser media, as its electromagnetic fields are compressed. Over long distances, the compression results in a measurable difference in propagation time. Interstellar gases, intergalactic plasma streams, all would have this effect that slows the propagation of light, while nothing that we know of affects the propagation of neutrinos.

The time-gap between the arrival on Earth of the neutrino burst, and the light burst, is surprisingly small if one considers that the light from the supernova had been propagated for 168,000 years. The light originated in a nearby galaxy named ' the Large Magellanic Cloud.' It evidently encountered a wide range of denser media on the way.

If the explosion theory was true, the time-difference would have been vastly larger. The internal fusion-sun theory states that it takes photons, the carriers of light, more than 10,000 years, up to 170,000 years, to make their way to the surface of a star, from the core, because of the volume of obstacles along the way. With this in mind, it takes a big leap of faith to accept that a shockwave can rip through the same, more than a million kilometers of material, and create a burst of light that arrives on Earth only 3 hours delayed from the neutrino burst that is supposed to have started the shockwave.

A tell-tale image

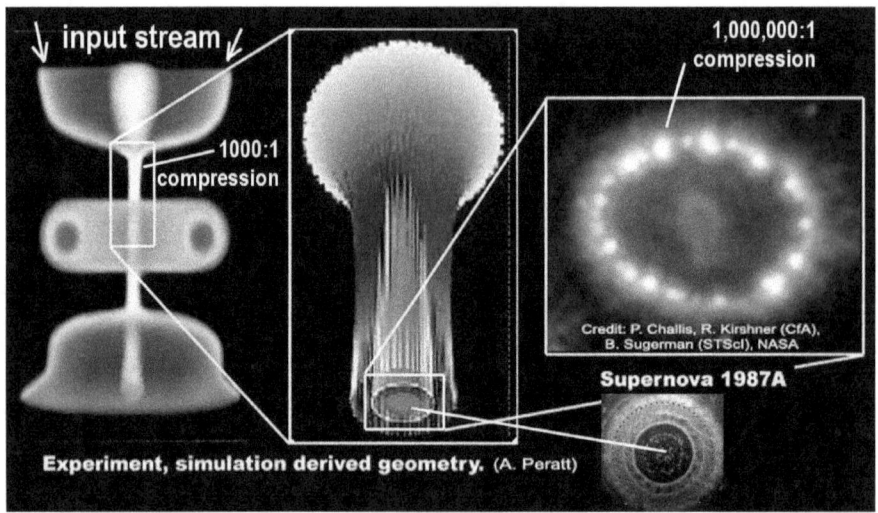

input stream

1000:1 compression

1,000,000:1 compression

Credit: P. Challis, R. Kirshner (CfA), B. Sugerman (STScI), NASA

Supernova 1987A

Experiment, simulation derived geometry. (A. Peratt)

However, the big proof that we live in a plasma universe is located in the ring of lights surrounding the nova event. The ring of lights is a tell-tale image that reveals the sheer size and intensity of the plasma currents in interstellar space. The currents became visible with the 'smoke' from the synthesizing fusion getting in the way of the plasma current, by which the current becomes visible. High energy physics experiments have revealed that intense plasma currents form 56 filaments of currents, that later combine into 28 filaments, electromagnetically arranged into a tube with the cross section of a circle. We see a large scale example, here, existing in space, with a nova event located in the center of it. The ring of plasma filaments evidently pinched and provided the plasma current for the nova event at the very center of it. It is rare that one encounters such clearly visible evidence that supernova events are electric events.

Electric-universe supernova events

The Crab Nebula

NASA

#40

The Crab nebula resulted from an evidently still larger plasma event that left behind a lot of 'smoke' to become visible in the plasma current streams. The Crab nebula is for this reason one of the most studied objects in the sky, but it is studied from the false theoretical basis of it being an entropic object, which it cannot be.

Its start-up may have been seen in 1054 A.D. in the form of a supernova. Since in the electric-universe supernova events are nothing more than large plasma discharge events - a type of galactic 'lightening' - the events are typically short-lived, on average for a few months, but the large ones like the Crab tend to be maintained thereafter by the plasma streams that had originally caused them in the first place. The tangle of the Crab's atomic filaments is actively maintained at temperatures in the range between 11,000 and 18,000 degrees, by plasma flowing through the system. If the Crab was not actively maintained it would have gone dark long ago. In fact, it may be dark again already. The Crab is roughly 6,500 light years distant and measures slightly over 13 light years across. The light that is seen in the photograph was emitted long before civilization had begun on earth. We see an electric light show presented here, of large proportions, but nothing more than that. Still, what we see puzzles researchers.

The Crab a pulsar

Crab Nebula NGC 1952 (composite from Chandra, Hubble and Spitzer)

NASA

What is deemed the central star of the Crab nebula is located deep within the nebula itself. It emits pulses of radiation across the entire radiation spectrum, from gamma rays, to radio waves. The radiation is pulsating at a rate of 30.2 times per second. The Crab has been termed, a pulsar, for this feature.

A neutron star is physically impossible

The Crab Nebula Pulsar
pulses 30.2 times per second

#41

Because the Crab's central star is seen being located at the focal point of a circular cloud, it has been theorized in entropic cosmology that the Crab Pulsar is the collapsed remnant of a star explosion that has left in its wake a neutron sphere 28 to 30 km across, with a spin rate of 1812 revolutions per minute. It is said that its fast rotating magnetic field causes pulsating havoc in the surrounding gases.

The problem with this theory is, that neutrons cannot exist by themselves, much less exist as stars. When neutrons become split off from their bond with protons, they decay back into being protons and become electrically active again. This means that a neutron star is physically impossible, and this includes all the derivatives of the theory, such as black holes, and so on. Nor would a neutron star have a magnetic field. Neutrons are electrically neutral. Only the movement of electric currents forms magnetic fields. This means that the pulsar is something much simpler than it is made out to be - something that is actually physically possible.

A pulsating system that generates high-energy bursts

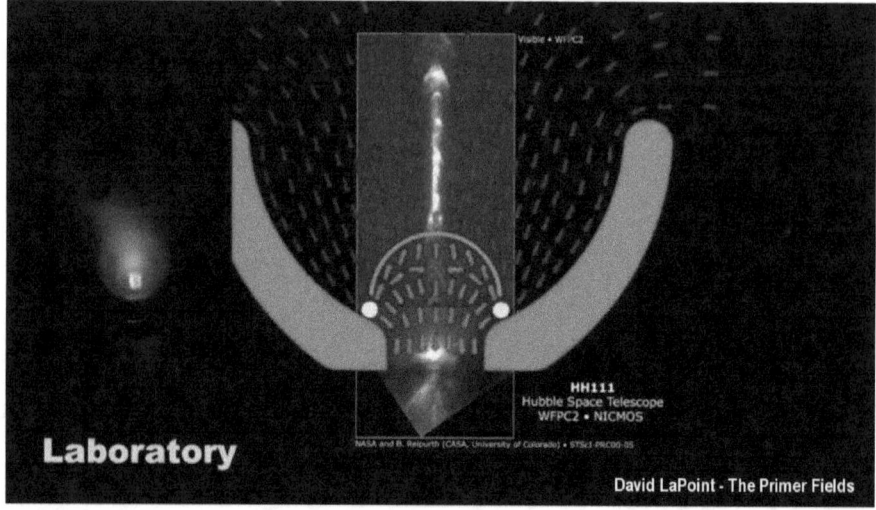

It is well recognized in plasma physics that when plasma is concentrated under the magnetic confinement dome of a Primer Field, and the plasma pressure is increased, a portion of the pressurized plasma breaches the magnetic confinement in a jet-like stream, by which the internal pressure weakens till the magnetic confinement shuts the flow off. At this point the pressure builds up again. The result is a pulsating system that generates high-energy bursts of concentrated plasma, that expands by the electric repulsion inherent in plasma.

The apparent star in the Crab

The Crab Nebula Pulsar
pulses 30.2 times per second

The circular cloud would be located within the plasma focus of the Primer Fields system. The apparent star in the Crab may not be an actual star then, but be merely the breakout point of concentrated plasma, through the confinement dome. This principle of plasma escaping the magnetic confinement dome of Primer Fields is evident everywhere. In our solar system, at the 'very small' level, we see it as the cause for the solar wind.

In the Eagle Nebula

"The Pillars of Creation"
Part of the Eagle Nebula

NASA,
Jeff Hester, and
Paul Scowen
(Arizona State University)

#42

On the cosmic scale, however, the giant Crab nebula is nevertheless still a relatively small nebula. In the Eagle Nebula the brightest star - of a cluster of 460 large stars - is all by itself a million times more luminous than our Sun.

And this star 'smokes' profusely as an atom-synthesizing nuclear fusion engine. The star is located behind the tip of the leftmost pillar of synthesized atomic material. The star is said to be 80 times as massive as our Sun.

The other pillars, too, have a large star located behind their top, likewise, which are similarly active. What you see here is only possible in an electrically powered universe.

Named the Pillars of Creation

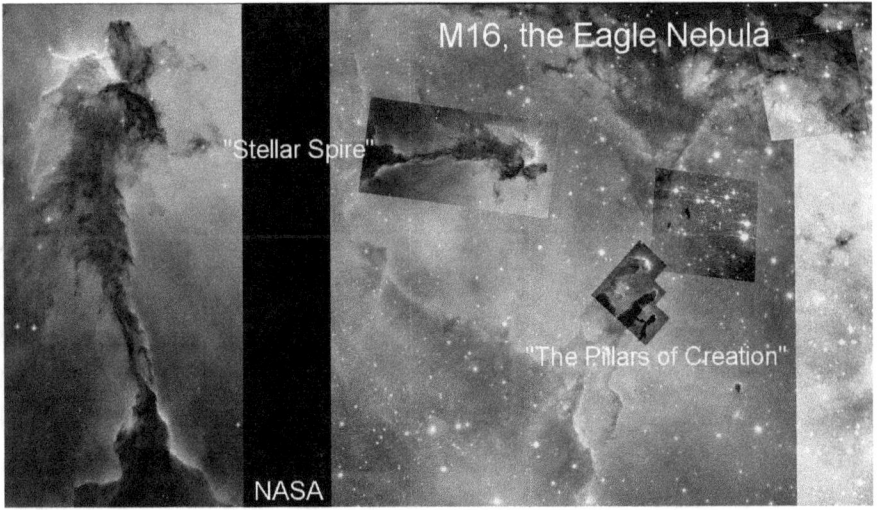

The three pillars themselves, named the Pillars of Creation, as huge as they seem with their million-times brighter sun behind each of them, are nevertheless but a small part of the larger sphere of the nebula that contains a number of extremely large features, including a large cluster of giant stars.

The gigantic features are barely recognized

And even the gigantic features are barely recognized when the nebula is seen as a whole, that is roughly 140 by 110 light years in size and is located roughly 7000 light-years distant.

At the focal point of a single large Primer Field

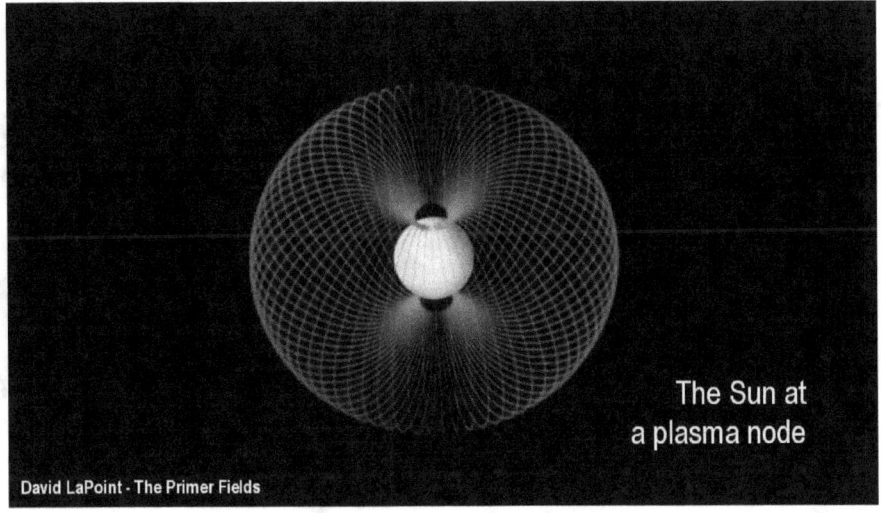

The Sun at
a plasma node

David LaPoint - The Primer Fields

The entire giant nebula may lay at the focal point of a single large Primer Field that powers the entire complex as a single unit, with multiple secondary fields operating at the local regions within.

In the electric universe

Milky Way look-alike
NGC 6744

In the electric universe, the operating principle is seen coming to light everywhere, and it is seen as being scaled upwards to almost any imaginable size, even to power an entire galaxy.

Intergalactic plasma streams

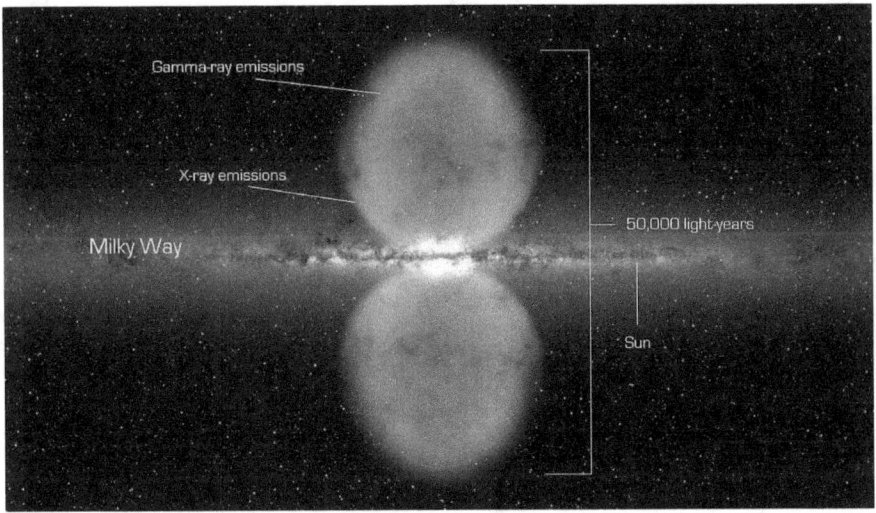

#43
The discovered plasma structures are the 'visible' confinement domes of the galactic size Primer Fields that concentrate and interface the intergalactic plasma streams that feed into and through the galaxy. The focusing of the Primer Fields creates a dense plasma halo around the entire galaxy.

Globular star clusters in the halo

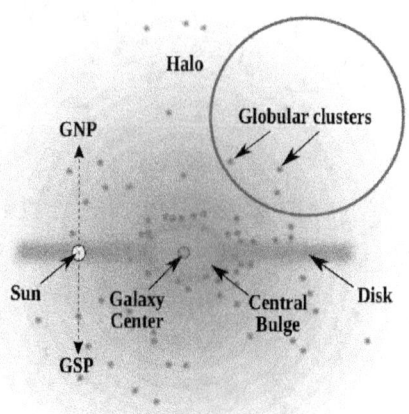

#44

The plasma concentration flowing through the halo is evidently dense enough to support a number of globular star clusters in the halo.

Spherical amalgamation of stars

Globular Cluster M10
(ESA/Hubble)

Star clusters are spherical amalgamation of stars that are typically up to 100-times more densely packed together than the stars in the neighbourhood of the Earth.

Globular star clusters

Globular Cluster M5
(ESA/Hubble)

The fact that the globular star clusters are almost completely free of dust and gases, can be seen as evidence that large networks of strong plasma currents flow through the halo region and sweeps up and disperses the solar synthesized atoms across the halo and the galaxy,

Intergalactic plasma streams resonance

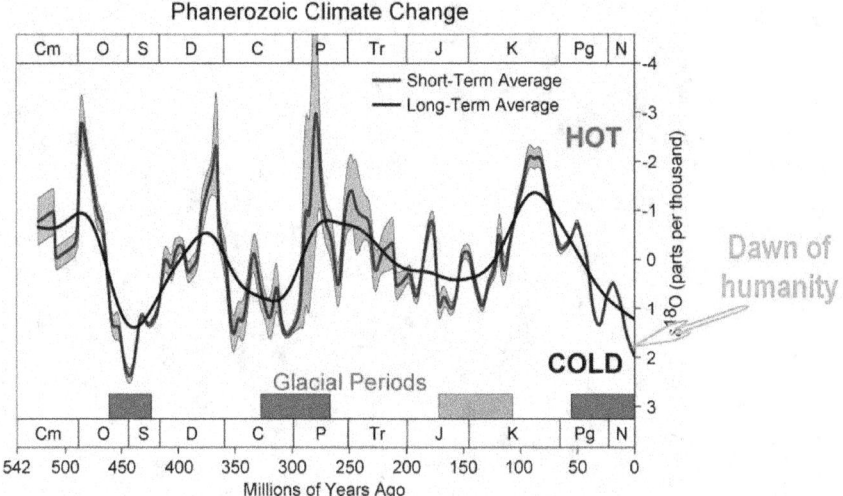

Phanerozoic Climate Change

#45

The intergalactic plasma streams that connect the Milky Way with the rest of the universe, have of course an electric resonance, each with a different resonance beat. We see the evidence of the two beats coming together reflected on Earth in the form of two long climate cycles overlaid on one-another. We see a long 144 million years, strongly dominant, climate cycle evident, which is modulated with a less dominant 62 million years' cycle.

The 62 million years' cycle is likely the plasma resonance cycle between the Milky Way and the Andromeda galaxy that is 2.5 million light-years distant.

The longer and more dominant climate cycle

galaxy Messier 83

NASA, ESA, and the Hubble Heritage Team

The longer and more dominant climate cycle is likely the resonance cycle between the Milky Way and the much larger M83 galaxy and its associated group of galaxies slightly over 10 million light-years distant.

The combination of two resonance effects

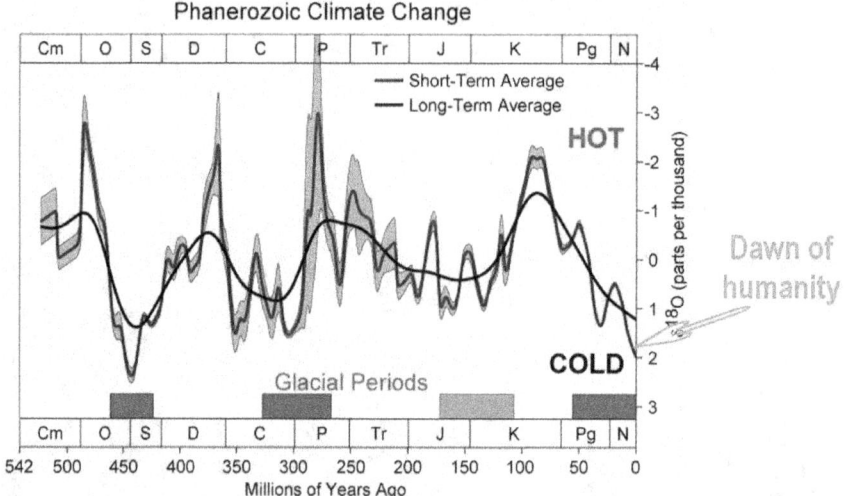

The combination of the two resonance effects has brought the entire Milky Way galaxy into an epoch of galaxy-wide weak electric conditions, which are presently at the weakest point of the last 450 million years. The weakening has become so severe that the modern epoch of the Ice Ages has begun, roughly two million years ago, which also coincides with the dawn of humanity.

The entire development of mankind

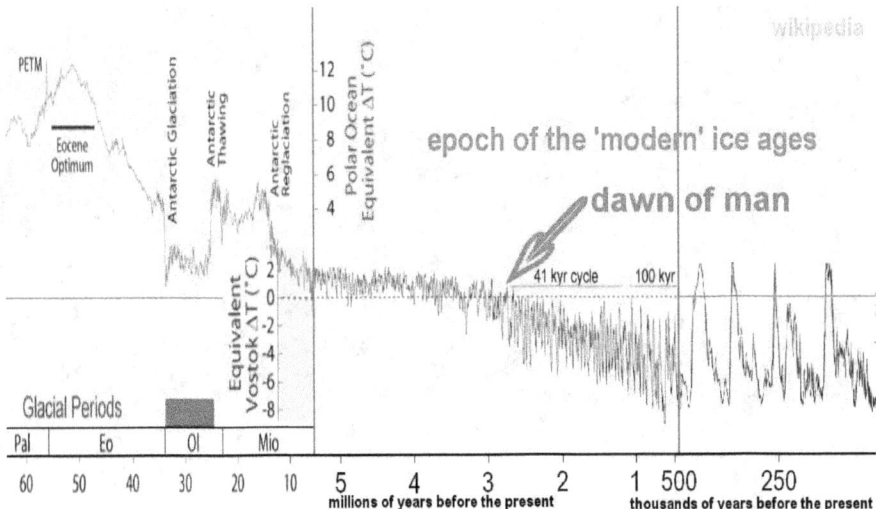

The entire development of mankind occurred during the epoch of the Ice Ages that consist of long glaciation cycles that get interrupted briefly with warm interglacial periods, like the one that we are presently in, which is about to end.

Our Sun is a rather mediocre one

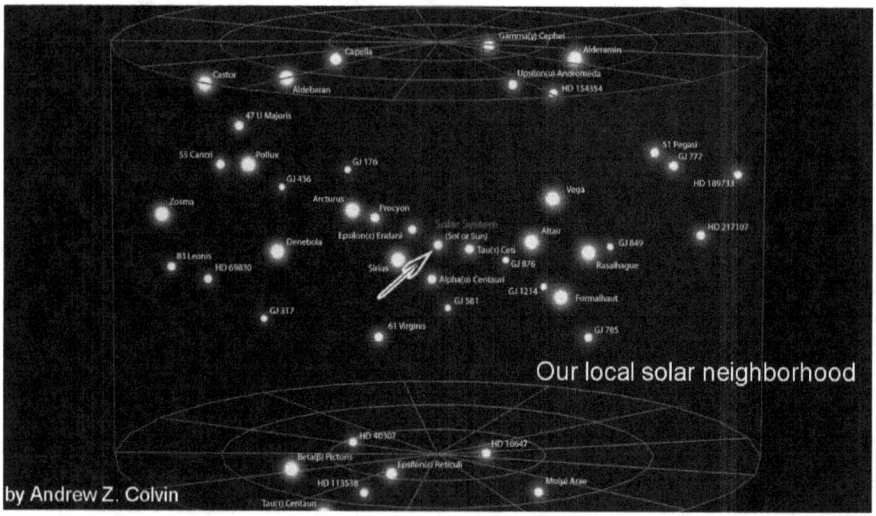

Our local solar neighborhood

by Andrew Z. Colvin

#46

The Earth is vulnerable, because our Sun is not a particularly strong star, nor is the solar system located in a particularly strong plasma environment.

Our Sun is a rather mediocre one among generally small stars that one finds located in thinly populated region of the galaxy where we are located. Evidence suggests that the electromagnetic Primer Fields that focus plasma onto our Sun are fast nearing the threshold point of collapse where the Sun goes inactive and the long-term normal glaciation climate begins anew.

Through the last glaciation period

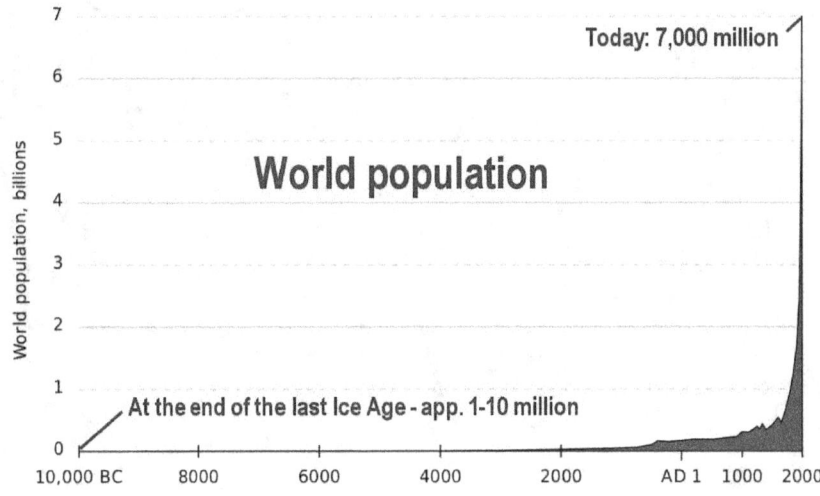

That the glaciation climate is a harsh climate is reflected in the fact that through the last glaciation period spanning 90,000 years, only 1 to 10 million people had managed to survive, worldwide. It took us a long time afterwards to recover and build up the man-created infrastructures for expanded human living. We now face the challenge to maintain ourselves as a large population through the next harsh period, and nor become devastated again.

Yes, 1 to 10 million people was the entire world population that existed at the end of the last Ice Age after more than two million years of the human journey. That's what we are getting back into, in potentially 30 years' time. That's the challenge.

Plasma streams between the stars Sirius and Vega

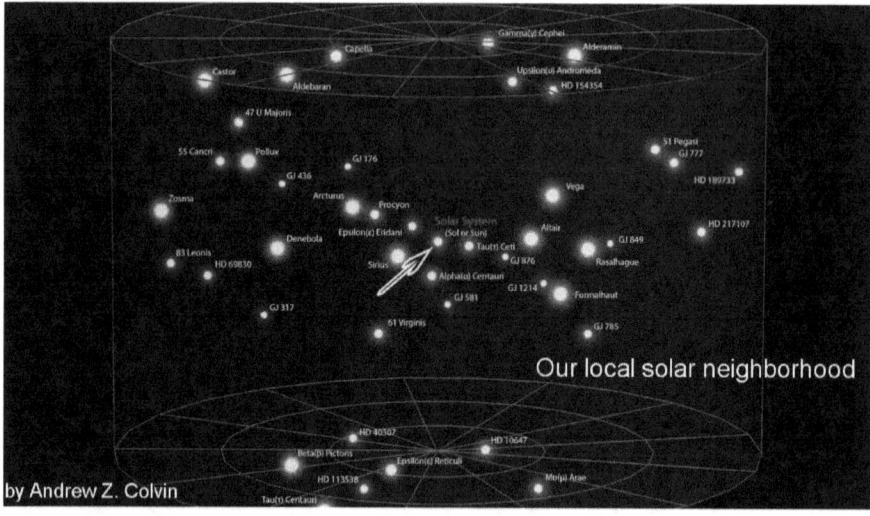

Our local solar neighborhood

by Andrew Z. Colvin

Our star, the Sun, may be more vulnerable than we wish to believe, whereby we are vulnerable too. This is likely so, because our solar system appears to be connected, electrically, to two somewhat smaller neighbour stars, named Epsilon Erandi and Tau Ceti, that are 10.5 and 12 light years distant, respectively. We may also be influenced by the stronger plasma streams between the stars Sirius and Vega, depending on which orientation is the most perpendicular to the planetary ecliptic. It is physically not possible to make exact determinations, based on measurements, because of the large distances involved.

Our Sun itself is not a strong giant either. It is one of the many mediocre stars in the local neighbourhood, and thereby one of the more vulnerable ones to the weakening electric conditions. Since interstellar plasma currents are inherently invisible, and too distant to be measured, the critical details of the interstellar electric connections that evidently do exist, are obviously open to a lot of speculation. We can at best only 'measure' their existence by their effects on the solar system, and thereby on the Earth. Fortunately, quite a number of these effects can be measured.

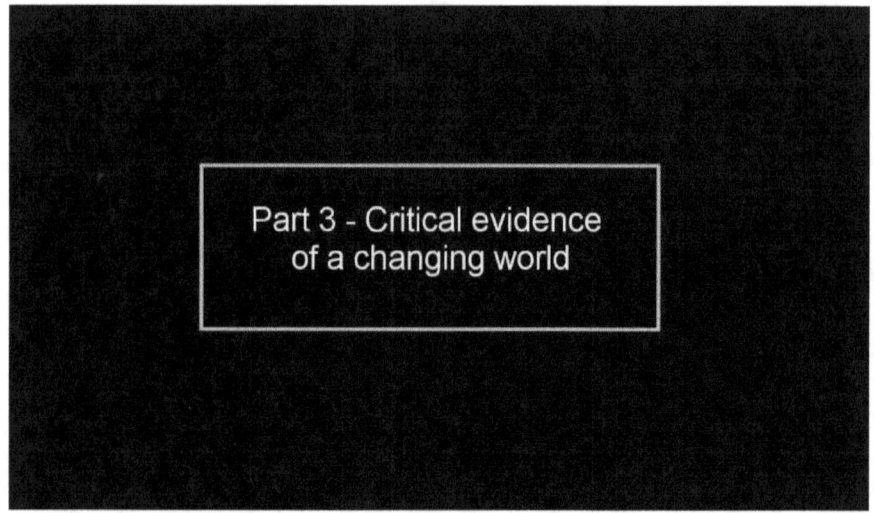

Part 3 - Critical evidence
of a changing world

Part 3 - Critical evidence of a changing world
Part 3 deals with the supporting evidence of the impending solar cut-off point. The evidence represents some rarely understood principles that render our sun a variable star.
Electromagnetic processes are variable processes. The intensity of the expression of the operating principles varies with the energy density in the driving processes. Let's quickly review the relevant principles that enable a Sun to be a variable star.

Dense plasma sphere surrounding the Sun

The Sun at
a plasma node

David LaPoint - The Primer Fields

The Sun can be seen as existing within a large electromagnetic
structure that concentrates interstellar plasma streams and focuses
them onto the Sun.
The results is a dense plasma sphere surrounding the Sun.

Primer Fields static experiments

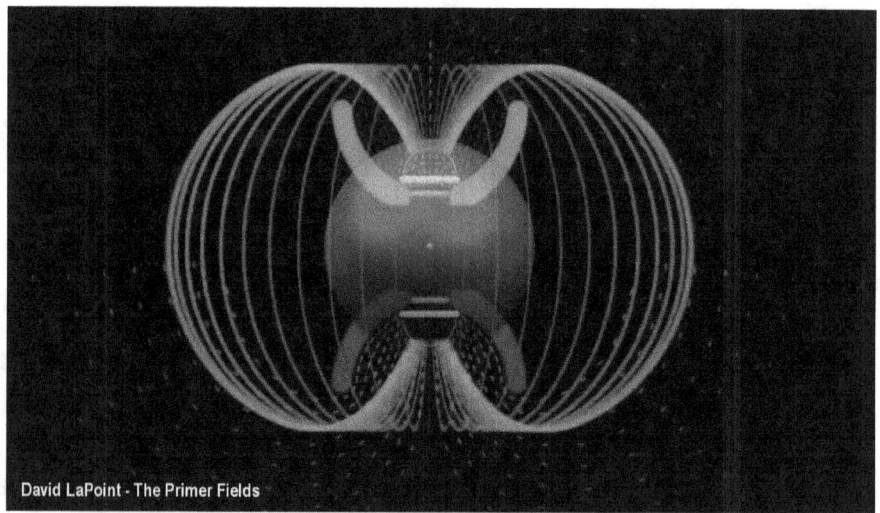

David LaPoint - The Primer Fields

As I said before, the dynamics of the electromagnetic structure have been explored in detail by the researcher David LaPoint, who has termed the structure, the Primer Fields. He explored the basic structure with static experiments.

High-energy electric flow experiments

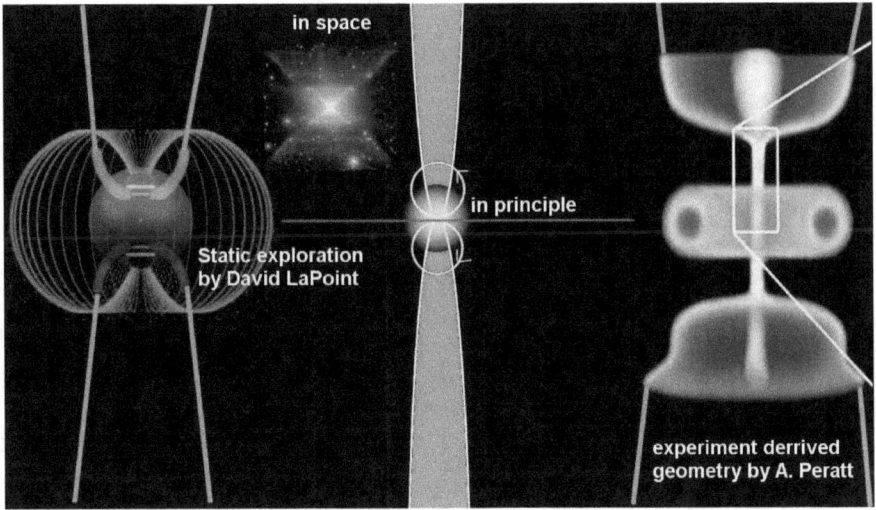

He used the static experiments to explore the details of the electromagnetic geometry that have previously been known and theorized from known principles, and have been seen evident in space, and were recognized also as a natural, universal, plasma-flow geometry as it has been observed in high-energy electric flow experiments, shown in the example on the right, produced at the Los Alamos National Laboratory presented by Anthony Paratt, the director of the experiments.

Revealed the existence of a magnetic confinement dome

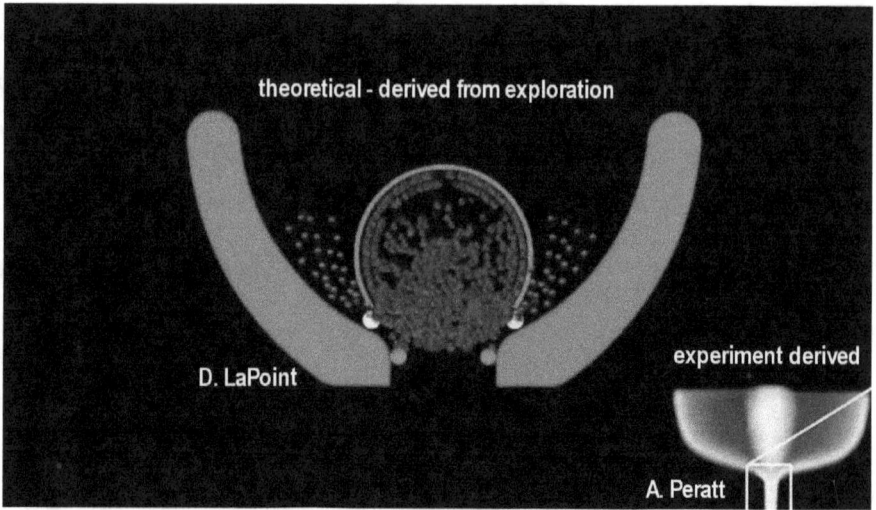

The static experiments have revealed the existence of a magnetic confinement dome and magnetic ring structures that facilitate the concentration of plasma.

Magnetic retention strength

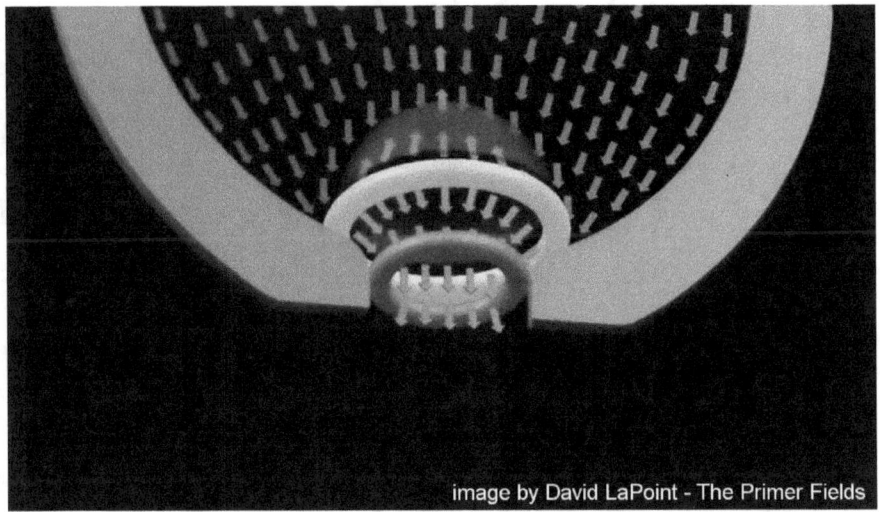

image by David LaPoint - The Primer Fields

It has been verified that by these natural features plasma becomes focused onto a sun.

David LaPoint also discovered that when the confined density exceeds the magnetic retention strength of the confinement dome, some of the excess escapes through the weakest point of the dome. By this feature that limits the plasma pressure that is focused onto the Sun, the Sun remains relatively steady.

Excess plasma becomes solar wind

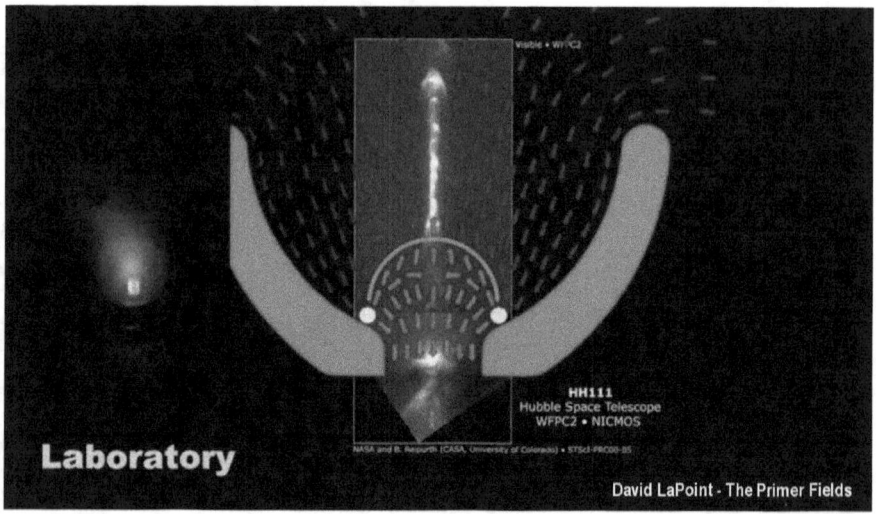

Excess plasma pressure becomes thereby simply vented back into space, typically in the form of solar wind.

venting' feature

This 'venting' feature too, has been replicated in laboratory experiments.

Concentrated in the fusion reaction cells

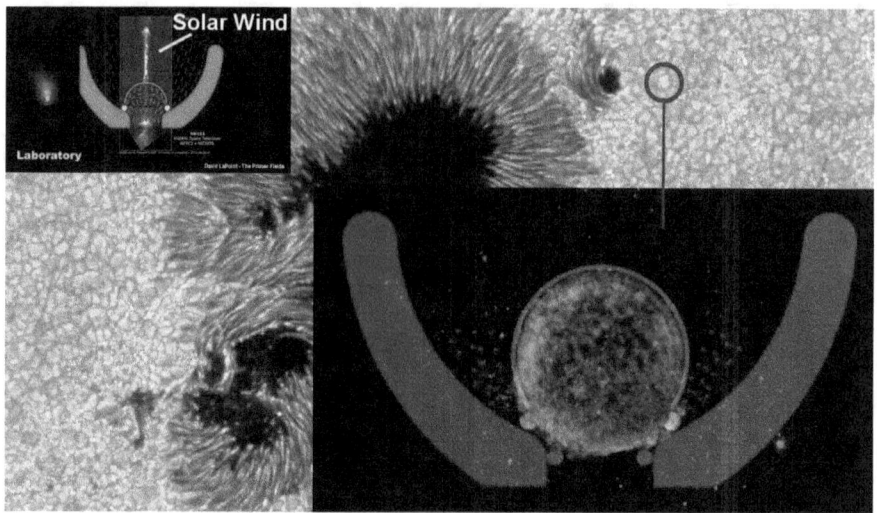

#47

The regulating principle applies especially to the processes on the surface of the Sun, where the highest concentration of plasma exists, which is forged from the concentrated plasma that surrounds the Sun, that becomes further concentrated in the fusion reaction cells to a density that enables nuclear fusion to occur.

Electrically powered fusion generates some thermal heating

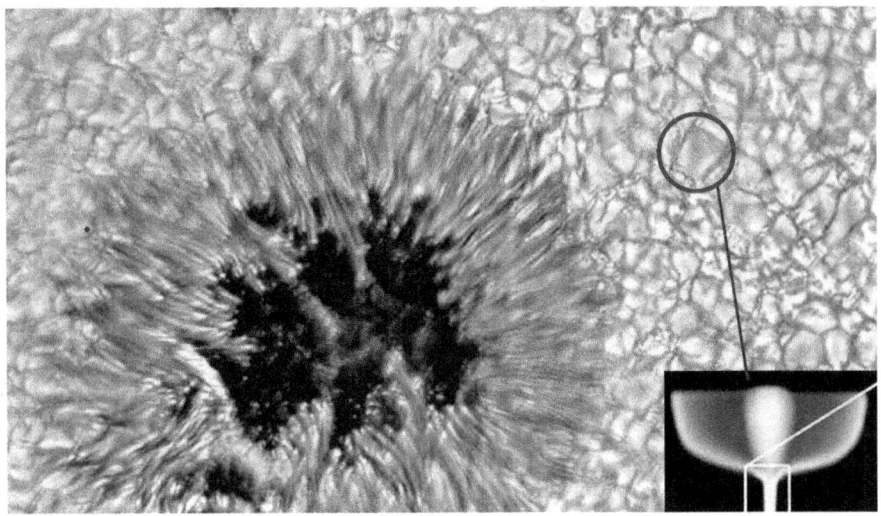

The resulting electrically powered fusion generates some thermal heating in the process, which heats the surface of the Sun to 5,505 degrees Celsius. When the input plasma has too high a density, which it currently still has, the excess plasma is ejected through the confinement dome.

For as long as the solar winds flow

© Milloslav Druckmuller/Barcroft

http://www.zam.fme.vutbr.cz/~druck/Eclipse/ - an example of the amazing solar eclipse photography
of Milloslav Druckmueller

#48

The ejected plasma from the confinement dome, flows away explosively from the surface of the Sun and becomes the solar wind. When the plasma's input density is extremely high, a strong solar wind results. Inversely, when the input density diminishes, the solar wind diminishes to a lesser wind-pressure. This means that for as long as the solar winds flow, we have enough plasma pressure focused onto the Sun to maintain the fusion process and with it the heating of the Sun.

The occurrence of sunspots is proof

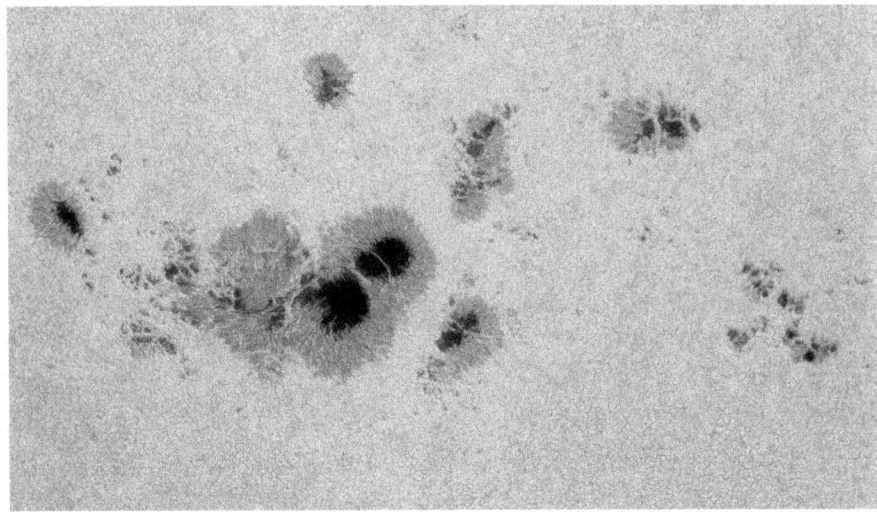

#49

Inversely, when the plasma flow is inhibited in the reaction cells on the surface of the Sun, such as by built-up localized back pressure, the electromagnetic processes that enable the reaction cells to function, breaks down. When this happens, the cells become inactive, either individual or in groups, even in large groups. The result gives us the sunspots. Sunspots occur when we have strong activity occurring that results in greater backpressure than the escape mechanism can handle. This means that the occurrence of sunspots is proof that we live under an electric Sun.

The occurrence of the sunspots also tells us that we have a strongly operating, 'healthy' Sun.

When the sunspots no longer occur, then we know that the plasma pressure onto the Sun has diminished below what is needed for optimal operation. That's cause to worry.

The measure of the solar-wind pressure

© Miloslav Druckmuller/Barcroft

http://www.zam.fme.vutbr.cz/~druck/Eclipse/ — an example of the amazing solar eclipse photography of Miloslav Druckmueller

#50

The most immediate measurement that we have for judging the health of the solar system, however, is not the existence of sunspots alone, but is the measure of the solar-wind pressure. When we see a trend of diminishing solar-wind pressure, extended over time, we have cause to become greatly concerned. This, unfortunately is the current state in the solar system.

NASA's Ulysses measured a 30% reduction

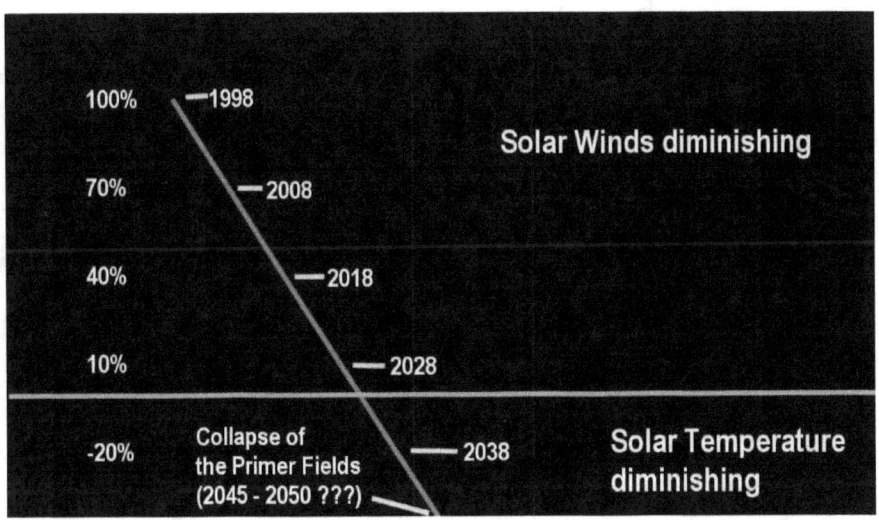

#51

The solar wind pressure has been diminishing for some time. NASA's Ulysses satellite has seen a brief window of this trend when it measured a 30% reduction of the solar-wind pressure in the space of ten years. This is a huge reduction. If this trend is projected forward in a linear manner, the solar wind will cease completely in the 2030s timeframe. Past the point of zero solar wind, any reduced plasma input pressure must result in a weaker Sun, or the Sun going inactive from a point on. We are not at this point yet. But we may get to this point soon.

In forecasting the future, we need to consider that the self-regulating system that keeps the Sun at a constant output level at the present time, can only maintain itself when excess input pressure exists. Once this is gone, anything can happen. The only unknown factor here, which no one can foretell, is, to what low level the solar output can diminish before the Primer Fields shut down that enable the entire solar process to function.

Magnetic measurements

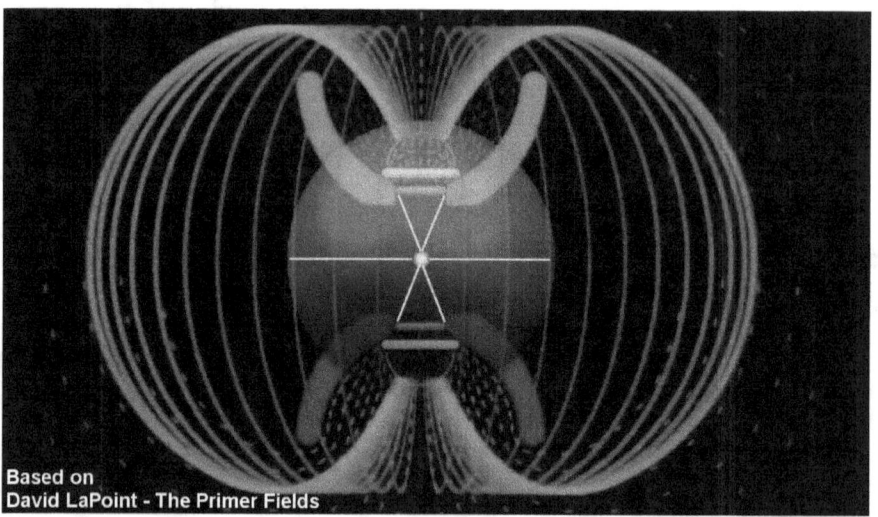

Based on
David LaPoint - The Primer Fields

Fortunately, other measurements are also possible that give us additional measurements as to where we stand. These are magnetic measurements.

It is physically possible to measure fairly accurately the 'health' of the 'primary' Primer fields that focus interstellar plasma onto the Sun.

The Primer Fields are essentially magnetic fields that are created by the self-concentrating principle of plasma flow, which generates bowl-type magnetic field structures near a focal point. These large magnetic fields that focus plasma around the Sun, interact with the Earth's magnetic field. The resulting interaction can be measured.

The dynamo effect of the spinning Earth

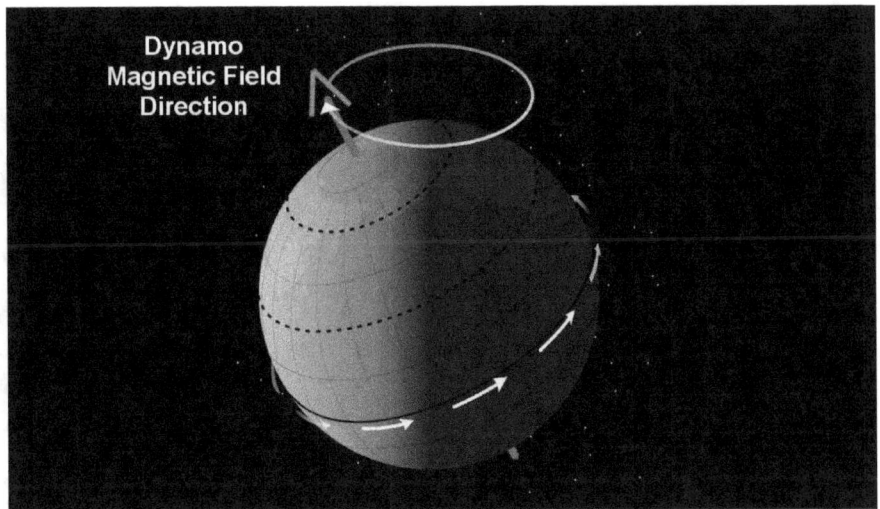

#52

The Earth's magnetic field is generated by the dynamo effect of the spinning Earth. Since the Earth has become pervaded with plasma particles that carry an electric charge, the spinning of the electric plasma creates a magnetic field perpendicular to the rotation, so that the resulting magnetic field becomes aligned with the planet's spin axis, regardless of how the spin axis is oriented. That's the dynamo-field effect. Since the Earth's spin axis is inclined 23.5 degrees off the ecliptic plane, the dynamo-magnetic-field is likewise so inclined.

Magnetic effect from the Primer Fields

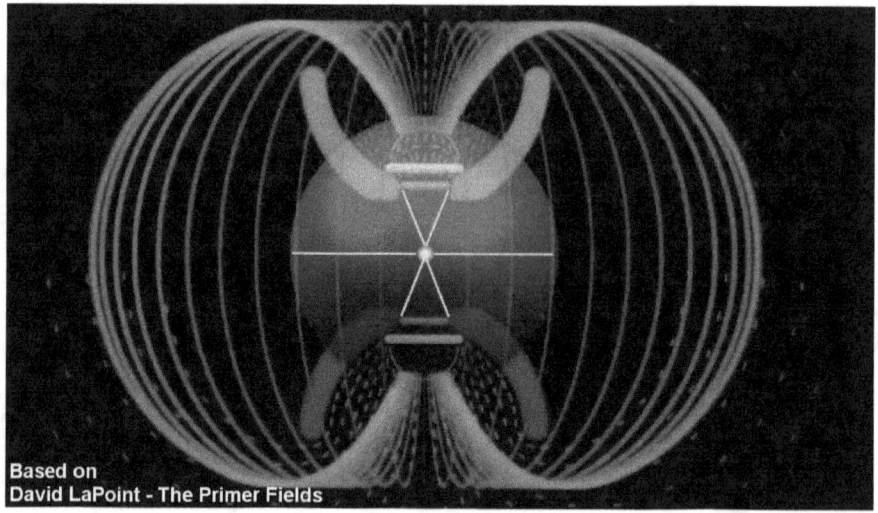

Based on
David LaPoint - The Primer Fields

But when the magnetic effect from the Primer Fields is interacting with the dynamo field, the resulting magnetic pole becomes deflected away from the geographic pole, towards the Primer Fields' magnetic orientation. The magnetic effect of the Primer Fields is oriented perpendicular to the ecliptic.

When the Primer Fields are strongly dominant

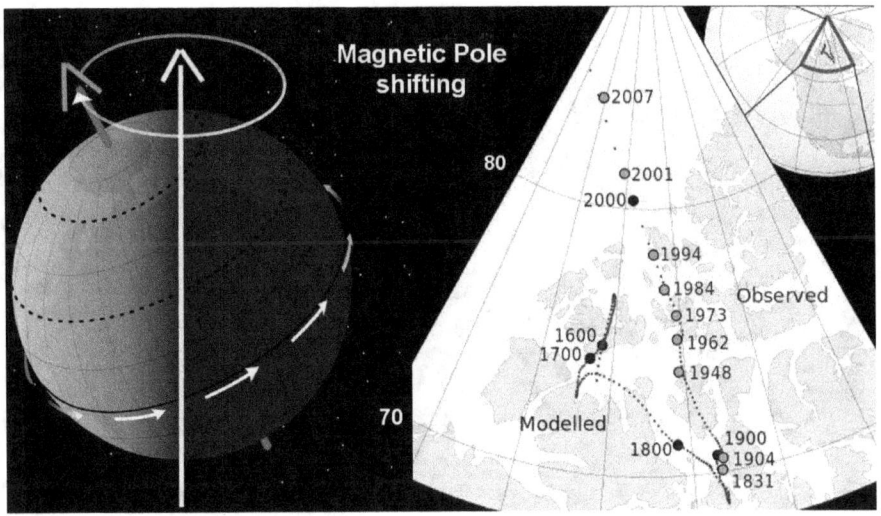

#53

All this means that when the Primer Fields are strongly dominant, the Earth's magnetic north pole becomes deflected away from the dynamo north pole, which is the geographic pole, towards the 70-degree latitude, in accord with the 23-degree inclination of the spin axis. This extreme condition existed in 1831, according to on-the-ground measurements conducted by an expedition in northern Canada. The marked location was the location where the effective magnetic field lines stood perpendicular to the horizon at the ground level where the measurements were made. Since the 1831 measurement was made, the electric environment in the solar system has become increasingly weaker.

Zero degrees' deflection would mean

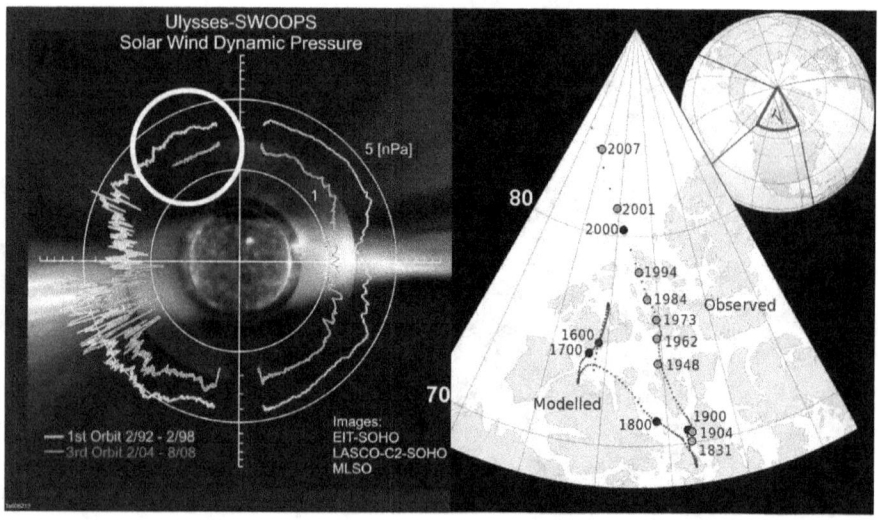

By the time NASA's Ulysses spacecraft began its measurements of the solar-wind pressure in 1992, the Primer Fields had already diminished so extensively that the deflection of the magnetic pole had already diminished so dramatically that barely more than half of the 1831 deflection had remained. By the time the Ulysses satellite was turned off, the magnetic deflection had been reduced to a mere 5 degrees. By 2014 it stood at only 4 degrees. Zero degrees' deflection would mean that the Primer Fields have diminished to zero, meaning that they would have ceased to exist. The magnetic measurements indicate that we are not far from this point.

A shift of the boundary zone

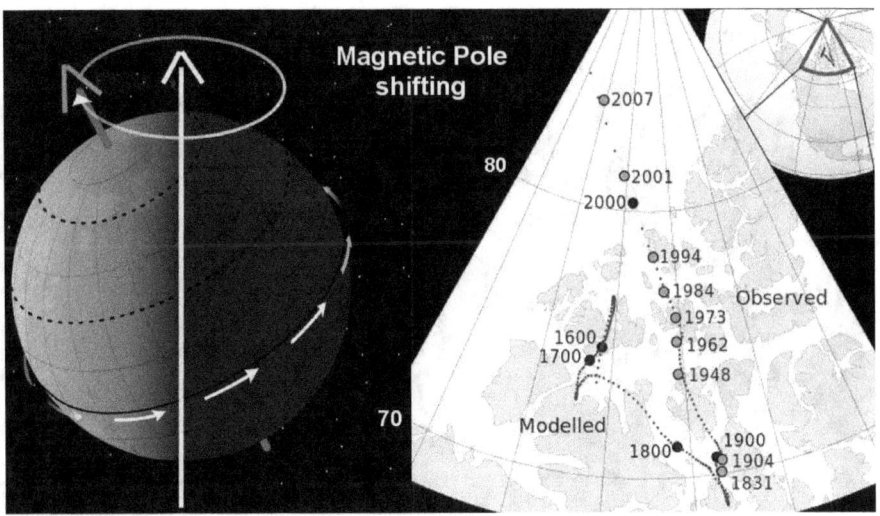

Of course, what we have measured is the combined effect of a number of principles interacting. The 'measured' magnetic weakening of the Primer Fields evidently also includes the effect of a shift of the boundary zone between the opposite magnetic fields of the in-flowing and out-flowing Primer Fields. The boundary would shift somewhat in a dynamically weakening system. This means that the shifting of the magnetic pole cannot be regarded as an absolute measurement. What is measured represents more likely a magnified effect. Nevertheless, the measured effect indicates that a tremendous weakening is in progress that has not been seen on Earth since such measurements were made.

Diminishing magnetic-pole deflection

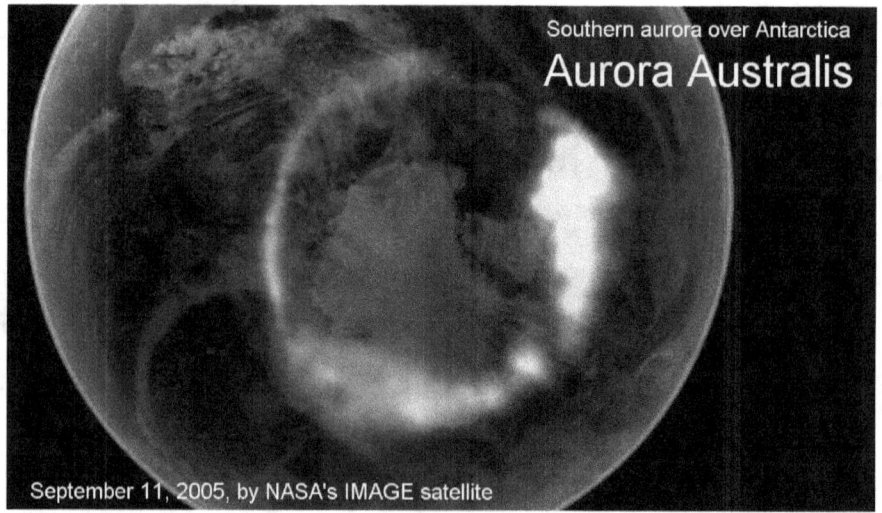

Southern aurora over Antarctica
Aurora Australis

September 11, 2005, by NASA's IMAGE satellite

The diminishing magnetic-pole deflection is now even visible from space. The location of the southern aurora confirms, that what has been measured in the North with magnetometers on the ground, is visibly evident from space. The southern aurora, as it is seen here from space, is barely 5 degrees offset from the geographic South pole. The reduced deflection is not insignificant if one considers that the overall magnetic field strength of the Earth itself, has diminished to possibly the lowest level since measurements were recorded. The total field-intensity of the Earth is evidently affected by the intensity of the all-pervading Primer Fields that may be far-more important for the Earth than we may know, which we are just beginning to discover.

Maintaining the spin-rate of the planets

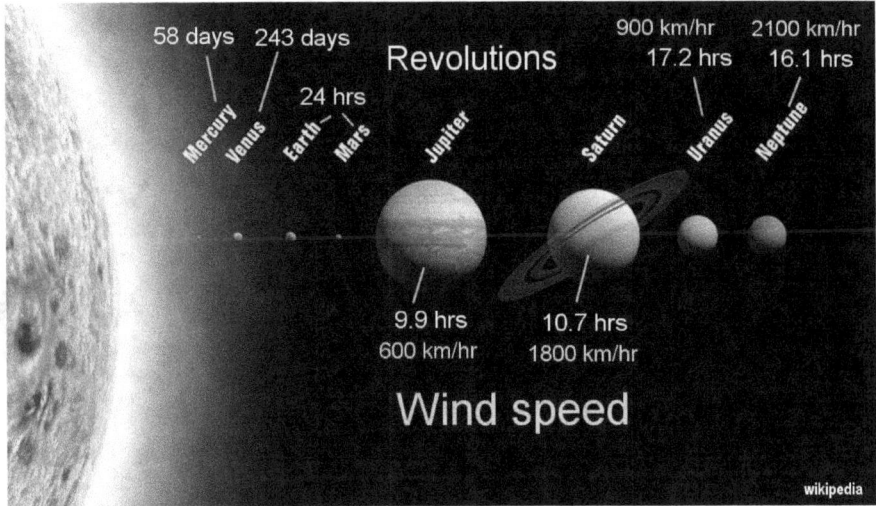

For example, it appears that the Primer Fields are critical for maintaining the spin-rate of the planets. If the spin-rate was not actively assisted, the planets would have stopped spinning long ago. It is interesting to note that the two innermost planets have almost no axial spin, which may be due to their being so close to the Sun that the swirling of the Sun's magnetic field overpowers the steady effect of the Primer Fields. While the Earth makes one rotation around its axis in 24 hours, it takes Venus 243 days for a single rotation, and less than 10 hours for Jupiter to accomplish the same. There is a pattern evident here, for which the cause remains yet to be discovered fully. When we get there, we will likely realize that many of our modern exotic theories such as curved space, time dilation, tunnelling effects, even aspects of quantum theory and relativity, and such exotic off-shoots as neutron stars, black holes, stellar explosions, and the Big Bang, may be nothing more than glorified epicycles of the type that Ptolemy has pioneered to proof a reality that doesn't exist.

Ptolemy had bent to the doctrine

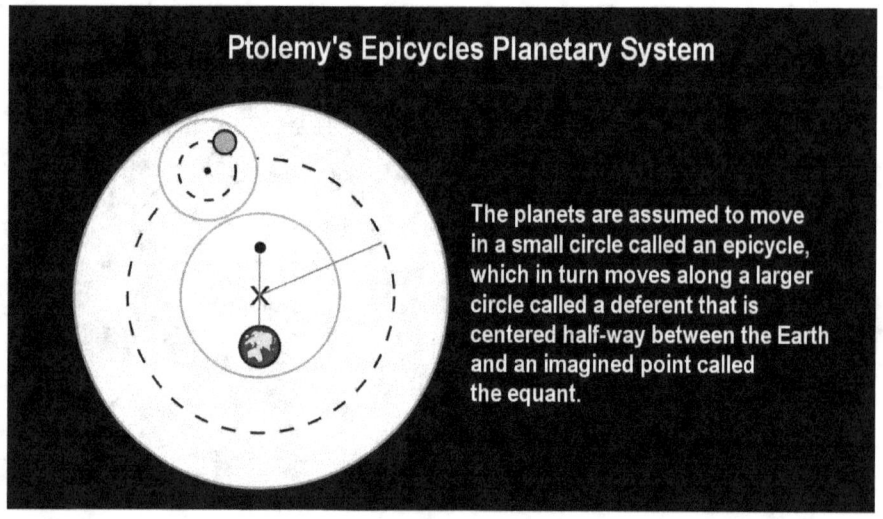

Ptolemy's Epicycles Planetary System

The planets are assumed to move in a small circle called an epicycle, which in turn moves along a larger circle called a deferent that is centered half-way between the Earth and an imagined point called the equant.

Ptolemy had bent to the doctrine that all heavenly bodies must follow the path of perfect circles, and that the Earth is the center of the Universe. This doctrine had ruled for more than a thousand years, and so did the epicycles theory that Ptolemy had invented in defending the doctrine. In today's age the ruling doctrine is, that mass and gravity are the only causative factors in the universe, so that all effects must be seen as rooted in their interaction. In the absence of the recognition of the electric nature of the universe and its vastly greater forces, extremely exotic epicycles have been invented to fill the gap, and to prove a 'reality' that doesn't actually exist, while ignoring an entire aspect of reality that actually does exist.

Hannes Alfven model of the electric universe

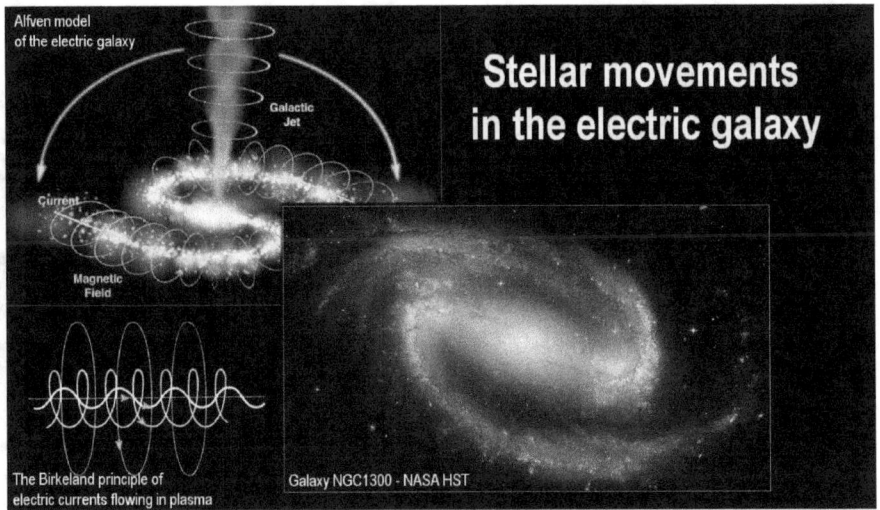

Some of this may be intentional. When Hannes Alfven developed his model of the electric universe in the mid 1900s, the Big Bang theory was developed as a counter pole. While the difference was academic then, as it didn't have immense worldwide consequences, those consequences are coming to light now and demand that the paradox that results from such differences, be resolved.

How soon the Sun will go inactive

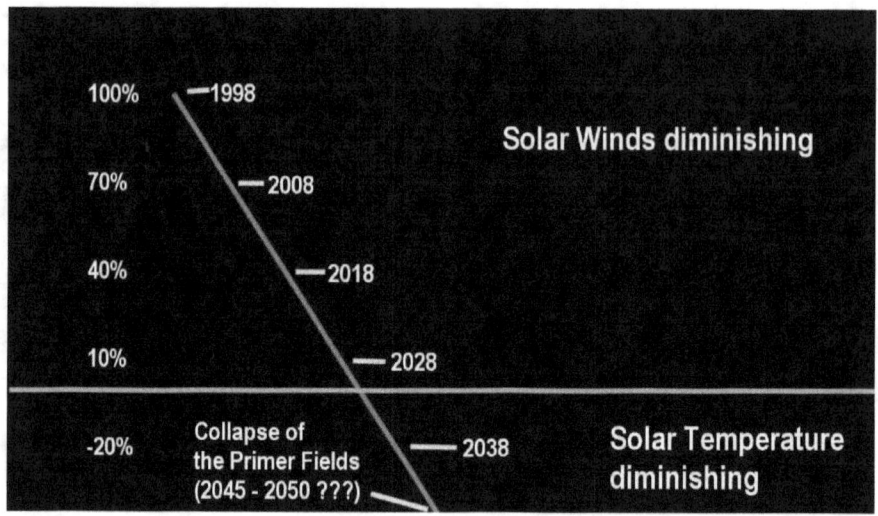

The determination as to how soon the Primer Fields will diminish to the point that the solar winds stop completely, and how soon the Sun will go inactive afterwards, is critical for the whole of humanity, though it cannot be forecast to the day and year. More factors than we may know, may determine this outcome. We can only recognize the principles that are glaringly evident, and the changing effects they cause that we have measured. This is all that we have available to respond to. But do we really need to know the precise day when the solar winds will stop, and how many days thereafter the Sun will go inactive? It is not the task of scientific forecasting to make such predictions. The task of science is fulfilled when it brings the principles to the foreground that can enable us to recognize with a great degree of certainty, on the basis of these measurements that we have already made, what type of future we are heading into, and what preparations must be made for humanity to live and prosper, and develop further, in that future.

Absence of tell-tale boundary conditions

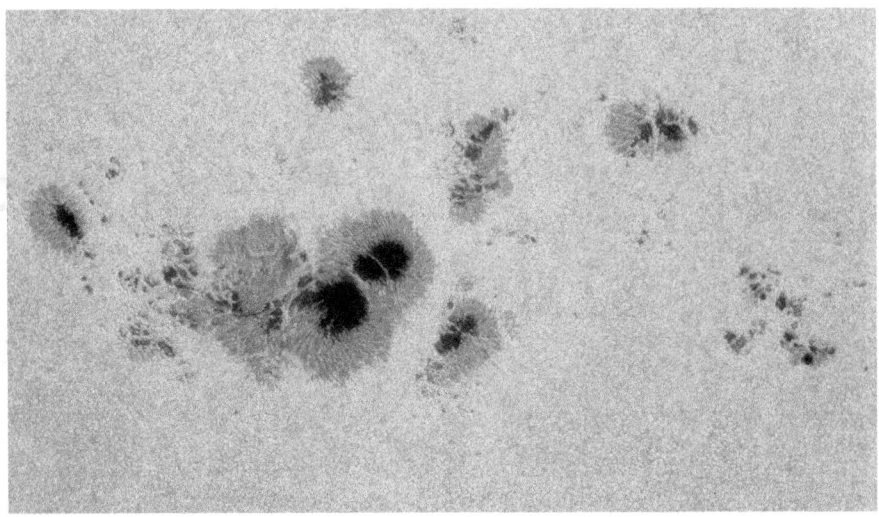

#54

That the Primer Fields as a dynamic system are highly sensitive to the prevailing electric conditions is evident on the Sun by the complete absence of any tell-tale boundary conditions that would indicate with certainty where the next sunspot will erupt. Until the very moment when a reaction cell 'blows out' and becomes a dark spot on the Sun, everything looks perfectly normal on the surface. We only know that under present conditions the sunspots will happen on the Sun somewhere. We also know from observing the sunspots, that once a sunspot has erupted, it takes a long time for it to heal, especially when the eruption extends over a wide area. Forecasting the start of the next Ice Age is not much different. the current projections indicate that we may have still 30 years left to prepare ourselves. But those are just projections.

Vulnerable to a sudden collapse without warning

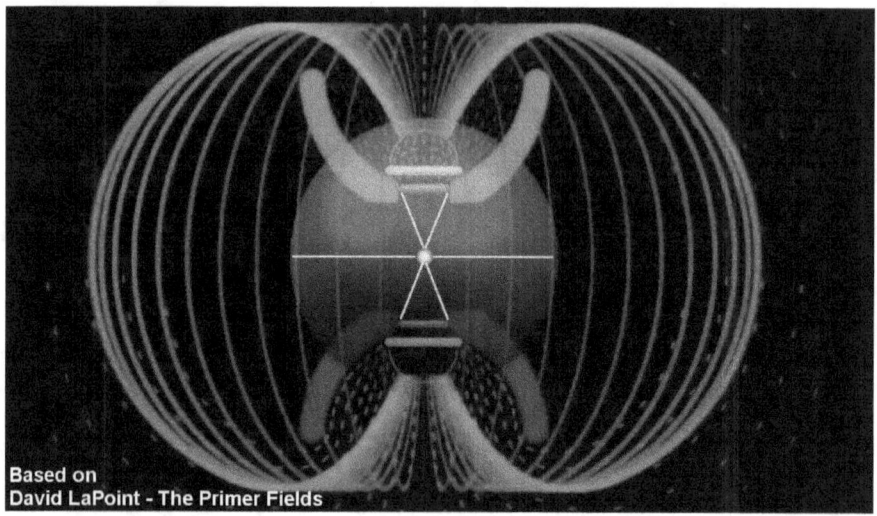

Based on
David LaPoint - The Primer Fields

The primary Primer Fields may be equally as vulnerable to a sudden collapse without warning, as the fusion-reaction cells are, on the surface of the Sun.

And when the primary Primer Fields collapse and render the Sun inactive, they may likely take a long time too, to heal. Ice core measurements indicate that this healing may take 1470 years, as it did throughout the last Ice Age, according to the measured Dansgaard Oeschger oscillations.

A reactive response is not sufficient

All this tells us that a reactive response to the unfolding event of the coming Ice Age transition is not sufficient as a measure for protecting humanity. The response has to begin now while we still have time to implement the protective measures based on understood principles and all the knowledge we have gained so far about them.

In this manner, by reacting before the crisis begins, we give ourselves a chance to survive and have a future. If we fail to act upon what we know, the principles we know tell us with a high degree of certainty, that we won't have a future at all.

Experience a 70% loss of solar energy

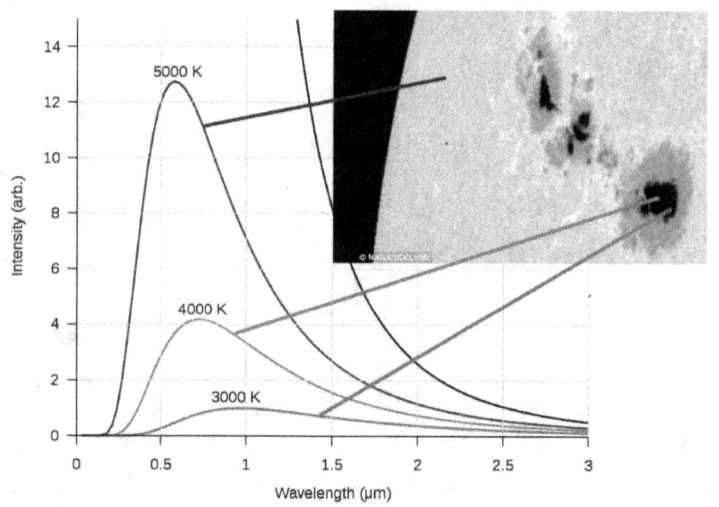

#55

When the Primer Fields collapse, the solar radiation will diminish rapidly to the low temperature that we presently measure in the umbra of sunspots, which appears to be in the range of 3,000 to 4,000 degrees Kelvin. This means we will experience a 70% loss of radiated solar energy the moment the Sun goes inactive. If we prepare ourselves in advance, we will live on in spite of the consequences. While the consequences will be harsh above the 40 degree latitudes, we can live with those consequences on the Earth quite easily if we have the requisite infrastructures in place. Unfortunately, the building of the type of infrastructures that are required, is not even envisioned today, much less set in motion.

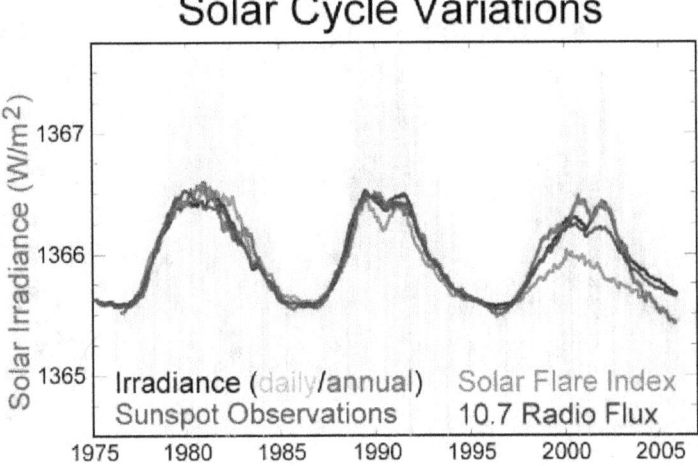

Solar Cycle Variations

The reason why the Sun remains at a constant energy level, even through its 11-year solar cycles, is that the Sun's Primer Fields system incorporates highly-efficient self-regulating features that keep the system operating efficiently under any circumstances, until the threshold is reached, beyond which the active fields collapse and nothing works anymore. We won't likely see a gradual change towards Ice Age conditions. We can only observe the insignificant-seeming fringe effects that we have observed, and react as if this is all the warning we get, which may be the case.

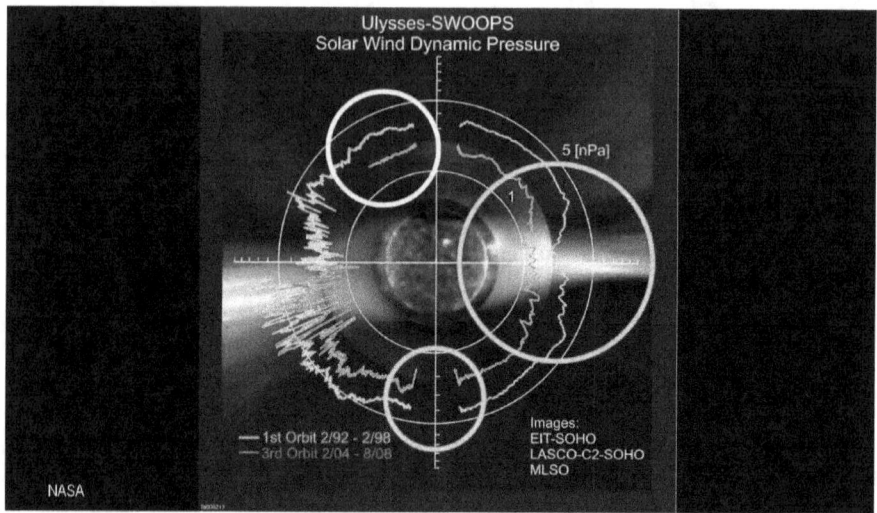

#56

This means that we need to look intensively at the fringe events that we know, like the diminishing solar-wind pressure that Ulysses has reported, in order to acknowledge the solar system as an electric system with a weakening electric environment, and that we do this long before the big threshold event occurs. The fringe-event measurements that were made by the Ulysses spacecraft, suggest that we may have still 30 years left to prepare our world before the greatest climate change in the history of humanity, begins. But will we do it? We see the huge diminishing of the solar-wind pressure, by 30% in a decade, and the reduced magnetic field of the Sun by the same 30% over the same time span. We also see the dramatic increase in Galactic Cosmic Ray flux by 20% as the result of the weakening.

Resulting from the increased cosmic-ray-flux

ISS-34 - Stratocumulus clouds

We see increased cloudiness resulting from the increased cosmic-ray-flux, which results in the cooling of the Earth as more solar energy is reflected back into space. We see greater temperature fluctuations now happening as the increased cloud forming reduces the water vapor, and with it the moderating greenhouse effect that is caused mostly by water vapor. We see increasing drought and flooding happening as the water-retention in the clouds is being reduced by the combined effect of the dynamic weakening on these numerous fronts. Surprisingly, society keeps its eyes and minds closed to the already visible fringe effects that are but precursors in a trend that promises to become increasingly more severe of the next 30 years, with immense consequences for agriculture.

We may not even have 30 years left

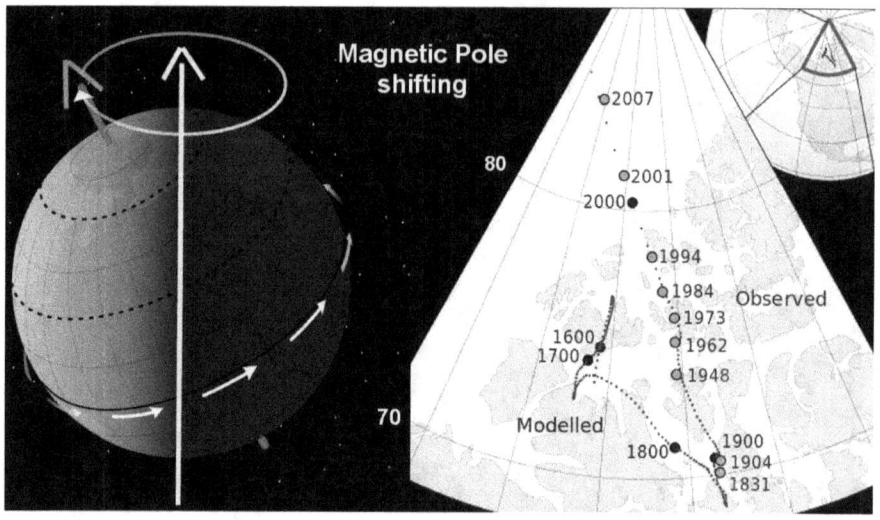

The magnetic measurements suggest that we may not even have 30 years left. But does anyone raise an eyebrow in the face of this ever-growing evidence? No! Instead, the world grinds on.

Electric weakening far-reaching reflections

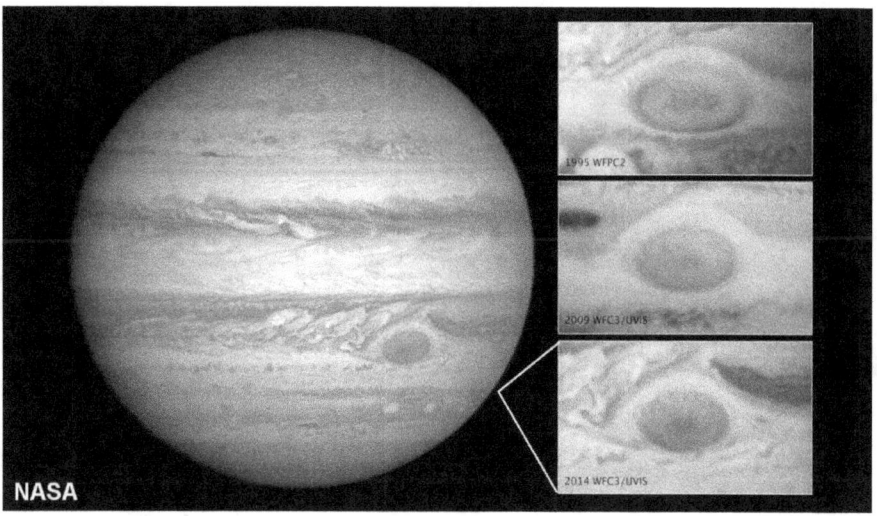

#57

The electric weakening that presently affects the solar system, has far-reaching reflections throughout the entire solar system. Some of these, we are just beginning to discover. One of the items of evidence that is coming to light is the weakening of the big red spot on Jupiter that has stood unchanged since it was discovered, which suddenly has become noticeably smaller over the last twenty years in which the major electric weakening of the solar system has accelerated. Do we have three more decades of time left? No one can answer this.

Might give us potentially 30 years

Ice Age of the dimming Sun in 30 years

www.ice-age-ahead-iaa.ca

Fortunately, we do have those known items of tell-tale evidence that a significant weakening is in progress, which, according to the principles involved, might give us a space of potentially 30 years, and maybe less, to prepare our world for living with an inactive Sun.

The proper response would be, to begin the work towards protecting our existence. The transition to an inactive Sun will likely occur rapidly, potentially within a day, or a few weeks, or a month at the most. The fast transition rules out reactive responses. The Sun may appear as a dark yellow star one morning that at first will glow with residual energy for some time. It will likely be larger in size, then, while the Sun's plasma sphere expands in the absence of external plasma pressure. It may also glow with some remaining plasma interaction.

White Dwarf examples

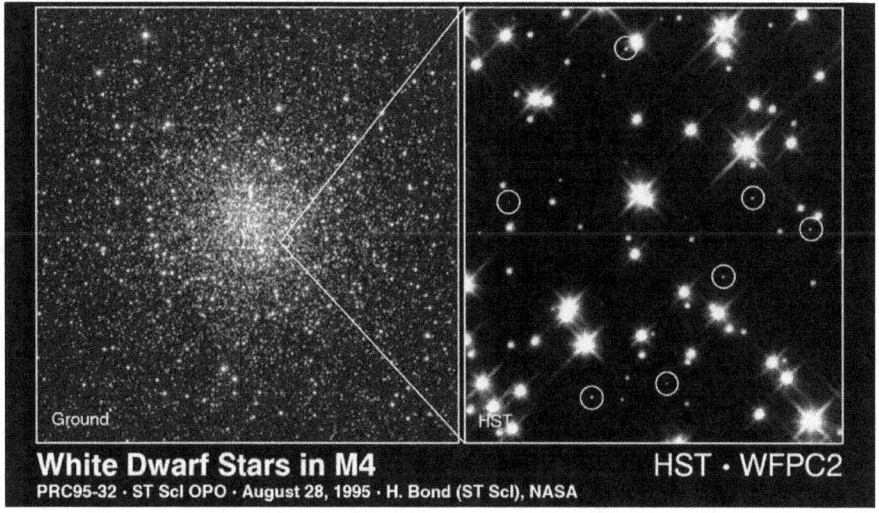

White Dwarf Stars in M4 HST · WFPC2
PRC95-32 · ST ScI OPO · August 28, 1995 · H. Bond (ST ScI), NASA

In time, however, the inactive Sun will likely cool down and
become a White Dwarf that shines from a small point in the center
with the energy of nuclear decay. A White Dwarf star, is a star that
is powered exclusively by the nuclear-fission decay of atoms that
are drawn to its core and are crushed there by the Sun's gravity.
Evidence for both stages appears to have been located by NASA's
Hubble Space Telescope in many places, like in the more-densely
populated star clusters shown here where such evidence would be
more likely detected. It is believed that up to 40,000 stars in the M4
cluster have gone dark. In the picture shown here, the White Dwarf
examples are encircled. Also examples of the dark red or yellow-
stage for stars are apparent. What we see here seems to tell us that
40,000 solar systems have already gone inactive in the cluster
shown here.

In the empty box of the entropic universe, a White Dwarf is deemed
to be a stellar remnant that is composed mostly of electron-
degenerate matter from a depleted star that has used up its fusible
fuel. Since the theoretical construct of electron-depleted matter

cannot exist, as electric repulsion would force the remnant apart and become infinitely diffused, the White Dwarfs are not what they are deemed to be. Since all stars are spheres of plasma, as they cannot exist as anything else, they cannot degenerate, burn out, or collapse.

A White Dwarf is a star that became inactive

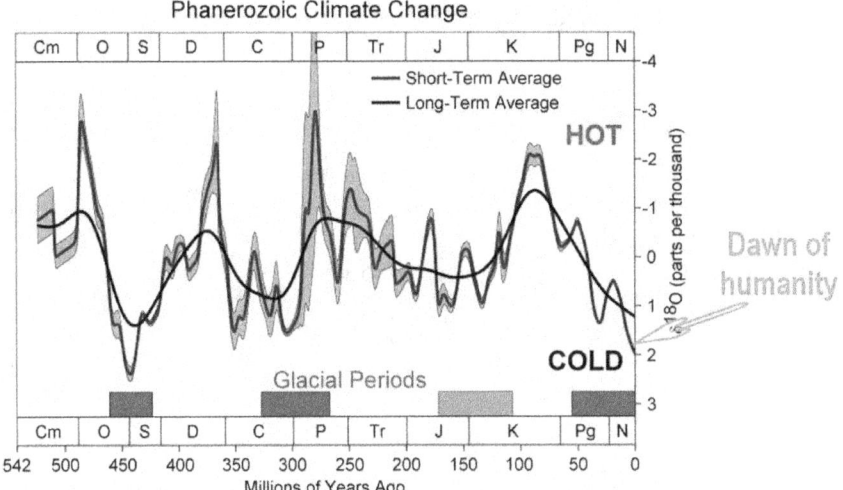

Consequently, a White Dwarf is a star that became inactive. In considering that the Milky Way galaxy has been in an extremely weakened state for the last 2 million years, it is not surprising that one finds a great many inactive stars in the galaxy. Most likely, all the inactive stars that we see today as white dwarves will become active again in due course, over time.

The Sun had recovered for brief periods

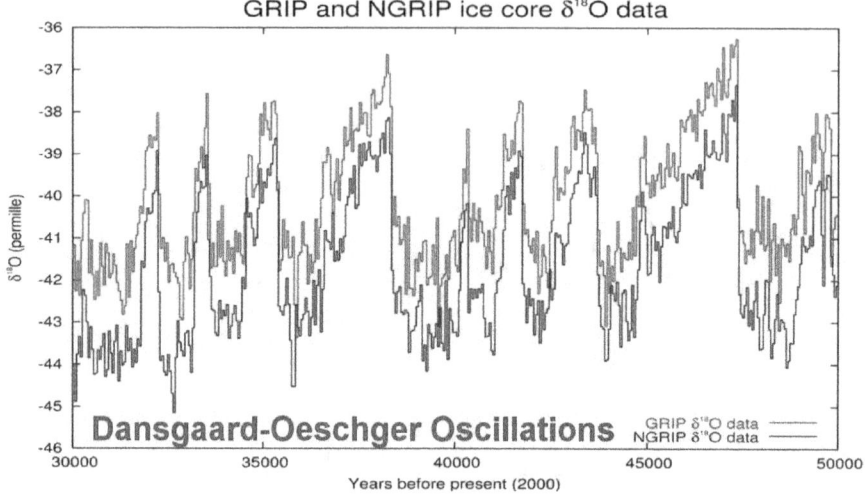

Ice core records from the last Ice Age tell us that the Sun had recovered for brief periods in beats of 1470 years, all the way through the glaciation period. The recovery would have repeatedly recharged the Sun's residual stores of energy and of atoms for its fission process while being inactive. Therefore, for most of the glaciation periods, our Sun would have joined the rank of the inactive stars.

Dust accumulations in glacial deposits

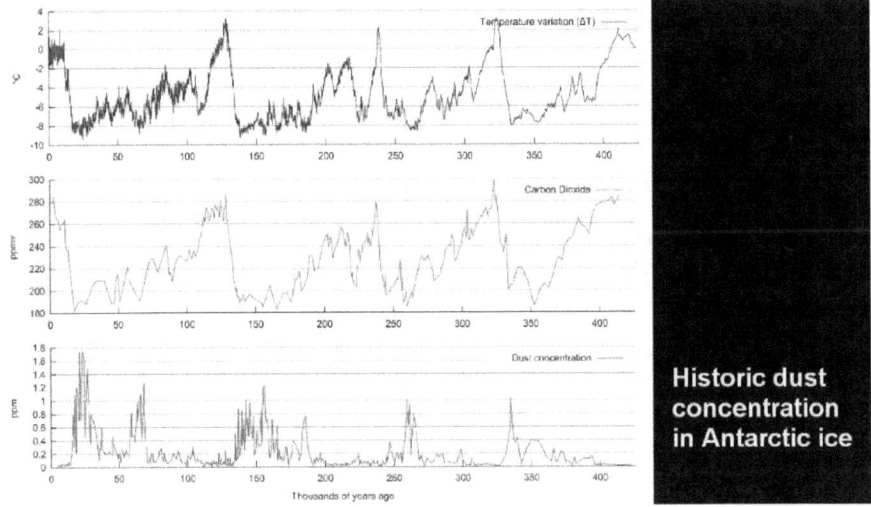

Historic dust concentration in Antarctic ice

#58

We do have ample evidence locked in ice on Antarctica, that the primary Primer Fields are real and have a larger effect on the solar system than people care to acknowledge. The dust in ice also tells us that the primary Primer Fields did not exist during the glaciation periods, and that as a consequence the Sun was essentially inactive during the glaciation periods.

The evidence exists in the form of dust accumulations in glacial deposits, near the end of every glaciation period that we have records for. The dust accumulation that we see is precisely what one would expect in times when the Sun has been inactive for long periods. In such a case, the electromagnetic assistance of the orbits of the planets would not have existed for long periods, which an active Primer Fields system brings to bear on the planets to stabilize their orbits.

The result of the loss of this effect would be that the orbits decay over time, especially that of the lighter asteroids in the asteroid belts. When the active maintenance for the orbits is missing for tens

189

of thousands of years, extensive asteroid intrusions would become normal on Earth, and would increase until the start of the next interglacial period, when the primary Primer Fields become active again. At this point the Sun becomes active again, and everything with it, and the orbits in the solar system become readjusted to normal again.

Thus, the ice on Antarctica tells us that the Primer Fields system is real and is powerfully effective when it is active, and that the ice ages are consequences of the Primer Fields being inactive.

Dust in ice, and White Dwarf stars

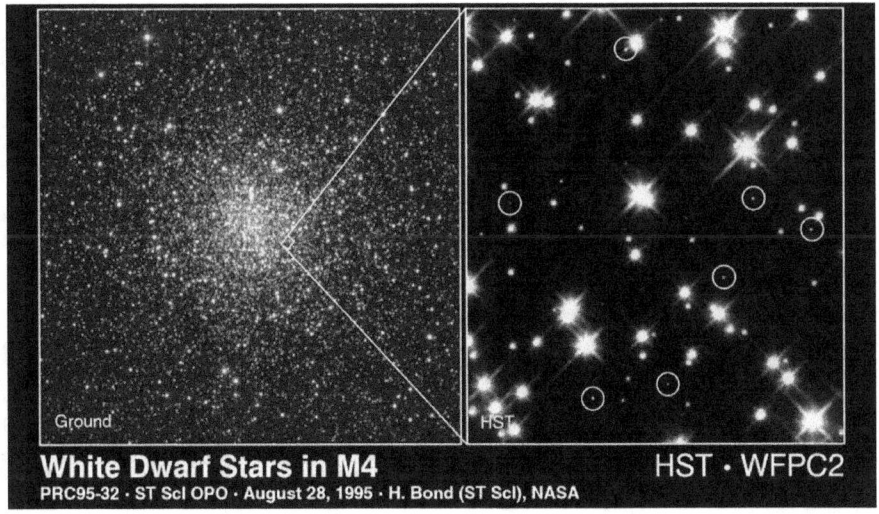

White Dwarf Stars in M4   HST · WFPC2
PRC95-32 · ST ScI OPO · August 28, 1995 · H. Bond (ST ScI), NASA

In the larger context, the evidence of dust in ice, and the existence of orange, red, and brown stars, even White Dwarf stars, all tell us the same story, each in its own dimension. And the story concludes - like the story of the solar wind and of the magnetic pole shift - that we must rouse ourselves to build the required infrastructures with which to protect our existence. This is the core reality that we cannot escape. We cannot run away from this battle and hope to live. We must win this battle. We are in a war, and the enemy is ourselves. The enemy is our small-minded thinking. It is called indifference, doubt, disinterest, inhumanity, a lack of love for one another, and so on. And this war must be won by all. The Ice Age challenge is not a physical challenge. The physical resources exist to meet the requirements for the impending climate transformation. But will we do it? The answer becomes a challenge rooted in the mind and involves the entire social, political, and economic scene.

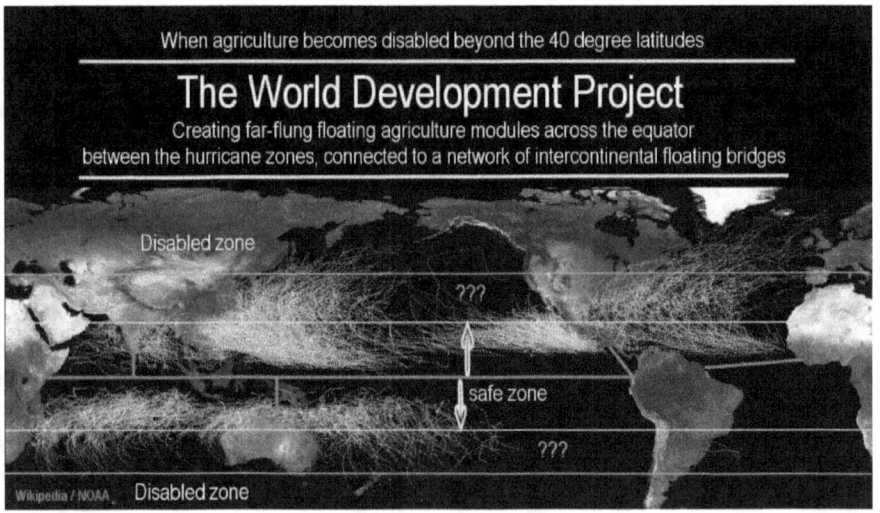

When the Sun goes inactive, the entire world is affected, and is extremely affected. The zone in which agriculture remains viable will shrink to a narrow band across the equator. Since little land exists there, most of the world's agriculture will have to be relocated onto floating platforms located across the equatorial seas, complete with floating cities, floating industries, and floating bridges to connect everything. As many as 6000 new cities will have to be built to enable the relocation of entire nations out of the by then uninhabitable zones, when the Sun goes inactive. While these necessary infrastructures can all be built, for which the material and energy resources exist in abundance, nothing is presently being build, or even considered. On this track we are presently loosing the war and are committing our children to death.

This failure is evident on the entire front, even in science. What is termed science, has been buried with perversion - buried under politically motivated doctrines, such as the manmade global warming doctrine where the truth, the discovery of principles, and raising the welfare of humanity, is no longer the objective of science. Science itself has been trashed to its very core by political perversions of it. And so, without a clear understanding of the actual principles that are expressed in the Ice Age dynamics, society will continue dreaming that it has a vastly longer period to prepare for the coming Ice Age transition, and that the severity will be negligible, instead of being extremely severe and being tangibly near. The point is that we don't have much time left to waste, considering the scope of the work that needs to be done. Ultimately the universe won't be cheated by us, no matter what we may choose for our future. If we choose to remain indifferent and get wiped off the map when the next Ice Age begins, a few may survive, and the resulting culture may stand at the same stage 120,000 years from now, that we stand on today and may choose more wisely and move forward. Only we, ourselves, are affected by how we choose. The potential certainly exists for the standard of truth to be raised once again, in science and in society, in the interest of the general welfare of humanity. When we start breaking ground on this front, then we have a basis for winning the war, and a basis for hope. Failure to move on this front is an act of committing suicide by default. Moving forward on this front is the path to life and freedom. We are in a race against ignorance, and this includes ignorance of what humanity really is and is capable of achieving.

Free universal housing in new cities

When this ground-braking progress begins, the human potential that we have, becomes realized, which is the potential we all have as human beings. Then, the awakening realization may inspire in society a commitment to break with its past of colonial, imperial looting and wars, even nuclear wars and doctrines for depopulation, and will inspire in society a commitment to the building of a renaissance of culture and freedom in human living, such as has never been imagined as being possible before. Even the concept of the nation and national culture will then become uplifted towards the common embrace of all mankind. If this uplifting of the identity of a nation as a human treasure remains unrealized, then many a great nation, especially from among the northern nations, will cease to exist when their territory becomes uninhabitable under the coming inactive Sun.

A nation isn't a land. It is far greater than that. A nation is a diamond, made great by productive and creative potential. The future rests on this potential. Land is only needed to fulfill the potential. If not enough land exists in areas where it needs to be, let's create it. Let's build vast automated industries for high-temperature processing, powered with nuclear energy, processing basalt for building the infrastructures with, and for the production of free universal housing in the new cities. If we commit ourselves to this, we are on course towards survival.

In honour of Canada's birthday celebration

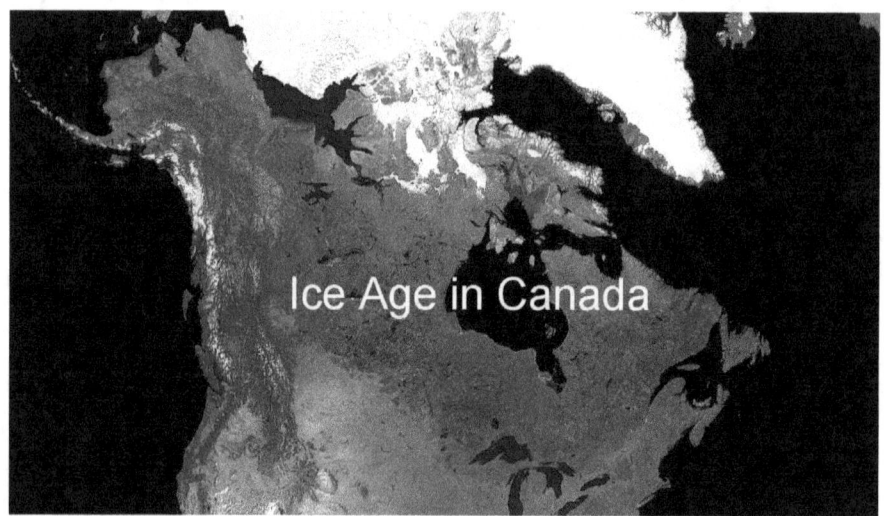

Canada, as a nation, can greatly contribute to the process. It owns the world's second-largest deposits of basalt, the needed energy resources to utilize the material, a commitment to technological progress, and a history devoted in many ways to the freedom of mankind. This video is produced in part in honour of Canada's birthday celebration, commemorating its formal founding on the First of July in 1867.

American birthday celebration on the Fourth of July

The video is also produced in honour of the American birthday celebration on the Fourth of July that commemorates the nation's founding with the adoption of its Declaration of Independence in 1776. These celebration days become meaningless when the nations' commitment to their future is lost.

Into which direction Canada and the world will be heading, will be decided in the near term wile time remains. The choice shouldn't be hard to make. Also it needs to be made by the whole of humanity. The failure to respond to the Ice Age imperatives would be synonymous with the world committing its children to death. Inversely, the timely responding to the imperatives would usher in a bright new world for everyone. And so, in considering what is at stake, I would say that we will make the effort to do what is necessary to survive, and not just to 'survive', but to have an endless future with abundantly rich living and happiness in every respect.

Supporting exploration videos

Supporting exploration videos by Rolf Witzsche

Celebrating the Near New Ice Age

Celebrating the Near New Ice Age

Cold Fusion powers the sun

Our Electric Cold Fusion Sun
(introduction)

NASA - view from ISS

Cold Fusion powers the sun

Ice Age of the dimmer sun in 30 years

Ice Age of the dimmer sun in 30 years

Our electric fusion Sun

Our electric fusion Sun

Electric planets electric Earth

Electric planets electric Earth

Electric Mars

Electric Earth - Electric Mars

Electric Mars

Energy future

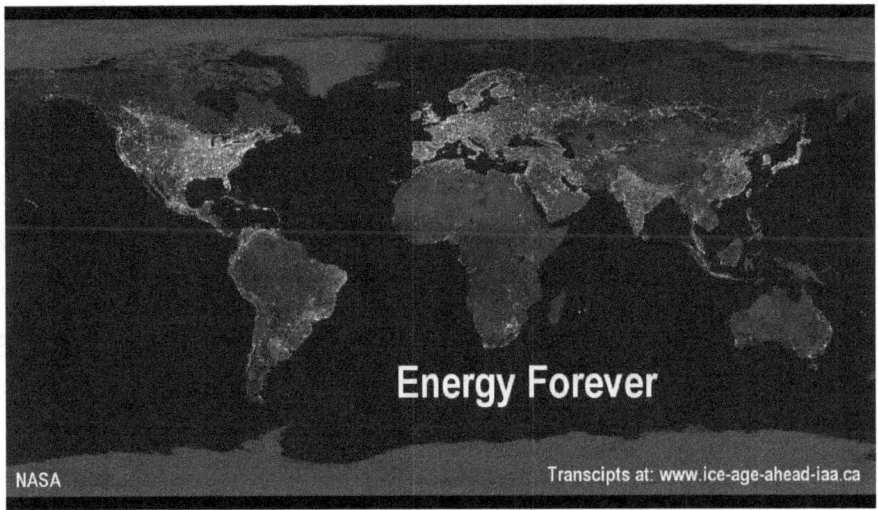

Energy future

Unlimited fresh water

Unlimited fresh water